BENGHAZI: KNOW THY ENEMY

by Sarah Adams and Dave Benton

A COLD CASE INVESTIGATION

Benghazi: Know Thy Enemy

Copyright © 2022 by Askari Media Group. All rights reserved.

All statements of fact, opinion, or analysis expressed are those of the authors and do not reflect the official positions or views of the U.S. Government. Nothing in the contents should be construed as asserting or implying U.S. Government authentication of information or endorsement of the author's views.

Note: The images of terrorists made available in this publication are from the author's archive. The images were personally derived during a decade-long, self-funded, open-source investigation into the attacks. Several images also were derived from social media scrapes and video exploitation analysis.

As the authors continue this cold case investigation into the U.S. Consulate in Benghazi and the CIA Annex attacks on September 11th and 12th, 2012, the public is welcome to provide information on any terrorists named in this book. Details can be emailed to: gazitips@proton.me.

ISBN 979-8-9868221-0-5 (Hardcover)
ISBN 979-8-9868221-1-2 (eBook)

Library of Congress Control Number: 2022916206

First Edition: October 2022

10 9 8 7 6 5 4 3 2 1

In Honor of Rone and Bub—May Justice Finally be Served

Contents

Prologue 7

Chapter 1: Al-Qa'ida's Benghazi Attack Planning 13

Chapter 2: Our Libyan Arrivals 21

Chapter 3: The Rising of the Black Flags 31

Chapter 4: Local Political Dynamics 39

Chapter 5: Enemies: The Masterminds 45

Chapter 6: Enemies: Power Players—Leadership and Commanders ... 65

Chapter 7: Enemies: Brothers in Arms—The Attackers 97

Chapter 8: Attack Network Analysis 243

Chapter 9: Our Final Thoughts 271

Acknowledgements 277

Attacks Timeline 281

Endnotes 307

About the Authors 317

PROLOGUE

This is not another Benghazi book. Here is an honest discussion about the facts regarding the specific terrorists who carried out the attacks, as available. The authors, two former Central Intelligence Agency (CIA) Officers, were intimately involved in the Libyan crisis before, during, and after the attacks. This book won't answer all your lingering questions like why the U.S. military was so ill-prepared to handle a crisis on 9/11 in North Africa, the hotbed of the world at the time, due to the unrest after the Arab Spring. Instead, herein is a personal account of our experiences having both served in Benghazi, Libya, and it's focused on our investigative efforts over the last decade to distinctly identify the attackers of the September 11th, 2012, attacks on the U.S. Consulate in Benghazi and the September 12th, 2012, attacks on the CIA Annex.

When you hear the word "Benghazi," the focus shifts to numerous inconsequential things besides who perpetrated these six separate attacks against our friends and us that fateful night in Eastern Libya. This book is written for all those who served in Libya, who have watched several individuals, for a variety of reasons, re-write history. This book is written for all our friends who finally want some closure and so our dear colleagues can rest in peace knowing that someone cared to identify the terrorists involved. This book sets aside the political pressure to downplay that the U.S. Consulate attack was planned and orchestrated by the Core Senior Leadership of al-Qa'ida.

In the following chapters, we will walk through our views on al-Qa'ida's plan to attack our Consulate after key senior Libyan leaders in al-Qa'ida Senior Leadership (AQSL) were killed in the Pakistan and Afghanistan border region as a result of U.S.-led drone strikes. We note that al-Qa'ida's former Leader, Dr. Ayman al-Zawahiri, as revenge, strategically decided to task the Leader of al-Qa'ida in the Lands of the Islamic Maghreb (AQIM), al-Qa'ida's most powerful affiliate in Africa, with the attack. This was because al-Qa'ida was still establishing its base in Libya, which was located in the Ganfouda neighborhood of Benghazi. AQIM co-opted their local allies primarily in Benghazi and Darnah, Libya, and then east into Cairo, Egypt, to help plan the attack logistics that led to the deadly U.S. Consulate in Benghazi attacks on September 11th, 2012, where we lost our Ambassador Chris Stevens and Department of State Information Management Officer Sean Smith.

On September 12th, 2012, a completely separate group, a local militia affiliated with the Libya Shield, carried out the deadly CIA Annex attack where we lost our close friends and CIA colleagues Tyrone "Rone" Woods and Glen "Bub" Doherty. In terms of Rone and Bub, that's how we communicated in CIA-designated warzones, usually by call signs. As persons may still be using prior call signs, none will be used in this publication except those of our deceased friends. As such, we will focus more on titles when it comes to our allies, but all attackers in the book will be in true name as they are the focus. There are approximately 50 different al-Qa'ida affiliates, additional global terrorists groups, regional battalions, and Libyan militias named in this publication.

What this book can tell you is what you need to know about the actual attackers in Benghazi, and the historic jihadist relationships among many of them as they fought a variety of international wars before coming together that night in Benghazi, be it in Sudan, Afghanistan, Algeria, Iraq, Mali, or Syria. Politics do not matter when the bullets are flying; teamwork, survival, brother and sisterhood, and sacrifice matter.

People want to know at the end of all this: Who did this and why? We answer who in here, we answer why in here, we share the status of those terrorists who have been brought to justice, but we leave a lot of questions on the table as to why the U.S. Government has done virtually nothing to bring these terrorists to justice. We ask that once you learn who are enemies are, that you help advocate for action to be done against the persons in these pages who are still at-large, and who are still a deadly risk to the U.S., its interests, and its allies.

This book will also discuss pertinent Libyan political dynamics, which did play a role in giving our "allies" in Benghazi a reason not to provide us forewarning and not to come to our defense during the multiple attacks. It frankly also gave them reason to participate themselves in the attacks. Look at this as a guide offering an overview of the terrorists who were residing in and traveling to Benghazi that fateful September, with the benefit of it being years after the dust has settled. It is time to learn, seek true justice, and close this chapter in history the right way—by going after the terrorists who got away with killing our Ambassador.

As a quick background so you know a little about who we the authors are, I was a counterterrorism analyst and targeter for almost a decade with the CIA, working in Libya starting in January 2012. My co-author Boon served as a security officer with the CIA for much of the same time, and he arrived in Libya in July 2012. Before and after the attacks, we worked in various capacities against al-Qa'ida. Boon had even helped in the Agency's tracking and locating efforts against al-Qa'ida's Founder, Osama Bin Laden, leading up to his death on May 2nd, 2011.

We both woke up at the CIA Annex on September 11th, 2012. It was an extra early morning as I needed a ride to Benghazi's International Airport. As I had a flight booked to Europe where I was to hold counterterrorism discussions with liaison partners regarding the threat from terrorists in North Africa. Boon was one of the Officers who drove me to the airport, and little did we know how the day would unfold as he stayed in the city throughout the attacks. We did not see each other again until 2013. After the attacks, he flew out of Tripoli, Libya on a relief flight late on September 12th, 2012, while I returned to Libya less than 24 hours after the attacks, and remained in country investigating the events until the end of November 2012.

We both took different paths after leaving Libya. Boon continued working for the CIA for several years and co-authored our CIA Global Response Staff's (GRS) first-hand account of the attacks as told in *13 Hours: The Inside Account of What Really Happened in Benghazi*, published September 14th, 2014. By that time, I had moved on to work on Iran issues at the CIA. Then approximately one month after *13 Hours* was released, I received a phone call from an unknown South Carolina number, and the caller identified himself as Congressman Trey Gowdy. I ended up chatting with Mr. Gowdy, the Chairman of the U.S. House of Representatives, Select Committee on Benghazi, and then Chairman of the House Oversight Committee, Congressman Jason Chaffetz. Even before the call, I was well aware that Mr. Gowdy had waded into the world of Benghazi investigations, as I spent much of my free time writing the names of Benghazi terrorists on a dry-erase board in my house. This fact will be quite evident as you flip through the "enemies" section in the middle of the book. After several conversations, he had convinced me, with some prodding from Boon, to leave the CIA and serve as the Senior Advisor for his Committee.

Taking a step back, it is important to note that Boon and I met several years before our work in Libya. I'd say it was a chance meeting, but Boon and another GRS officer (who I jokingly called "the other short GRS officer") were standing outside my door, knocking one evening. Assuming they were alerting me to something security-related, as in all my time in country, no one ever knocked on my door to say hello. I talked a little through the door before realizing they were inviting me over to watch a movie. Yes, sometimes the CIA is just that mundane. I agreed, as long as I could bring the movie I had just started, *Boondock Saints*.

I didn't know that circumstances would still have us working on the same mission more than a dozen years later. Like with many persons who have served overseas, you become friends and family for a lifetime. While we'd have many adventures overseas, both together and separate in the coming years, we first had to get through 2009. That year ended as an incredibly dark chapter for us in the CIA, and one of those reasons again was Zawahiri. It was a ruse to identify Zawahiri's location that led to the deaths of seven of our friends and close colleagues at CIA's Khowst Base, in Khowst, Afghanistan, on December 30th, 2009, by double agent and suicide bomber Humam Khalil Abu-Mulal al-Balawi with alias Abu Dujana al-Khurasani.

I remember when I learned of the Khowst attack. I had been traveling regionally, and upon returning, I was called into my manager's office. Even though the CIA was a "truth to power" organization, he was one of few bosses stuck on a very strict method of following the chain of command and usually berated me for going above his head for one reason or another. He was by the book, and I was rogue…at least as much as I could get away with. Quickly, I started to think of the list of things

I hadn't included him on over the last week so that I could mount my defense—that's when he immediately shut his office door and blurted out that he had bad news. I was shocked. Boon was back home in the States, so it was not until January 2010, when he arrived back in the country, that we were able to discuss the events that had occurred. He had lost three of his long-time colleagues as well, so the wounds were felt by us both.

Just two and a half years later, in June 2012, Zawahiri was incensed by the loss of his deputy, Abu Yahya al-Libi, with real name Mohamed Hassan Qaid, an Islamic scholar from Libya. Consequently, he directed a retaliatory attack against the United States' presence in Libya to honor Abu Yahya. Zawahiri made no secret of broadcasting his direct involvement in the impending U.S. attacks when, on September 10th, 2012, he issued a video statement calling for attacks on Americans in Libya to avenge the death of Abu Yahya.

Frankly, we ourselves had also been incensed by Zawahiri's continued targeting actions against our cadre at the CIA, and definitely tipped our hats to our former colleagues when on July 31st, 2022 they targeted Zawahiri in a deadly drone strike in Kabul, Afghanistan. Ironically, I had been working volunteer evacuations from Afghanistan for approximately 11 months prior to his death, and one of the Afghans I aid had a business right near the property where Zawahiri was killed. Ironically, he had to close his business in the arts industry when the Taliban took over in August 2021 as the majority of his employees were female. You never give terrorists an inch, or they will take it all.

In parting, Benghazi isn't just our story; it has become an American story, and it's time to share the facts regarding the al-Qa'ida terrorists who were responsible. In opening the files to our personal investigation into the attackers, we hope this book honors the memory of those we lost by showcasing our real enemies instead of allowing others to change the narrative to make us enemies of one another. Al-Qa'ida, its global affiliates, and its allies did not succeed in dimming the memory of our CIA brothers Rone and Bub, nor that of Sean, and will never diminish the hope Ambassador Stevens was to the city of Benghazi. At the end of the day, it may have taken a decade, but it's never too late to get the story right. Here we let you into our world, so you can know thy enemy just like we do.

CHAPTER 1

Al-Qa'ida's Benghazi Attack Planning

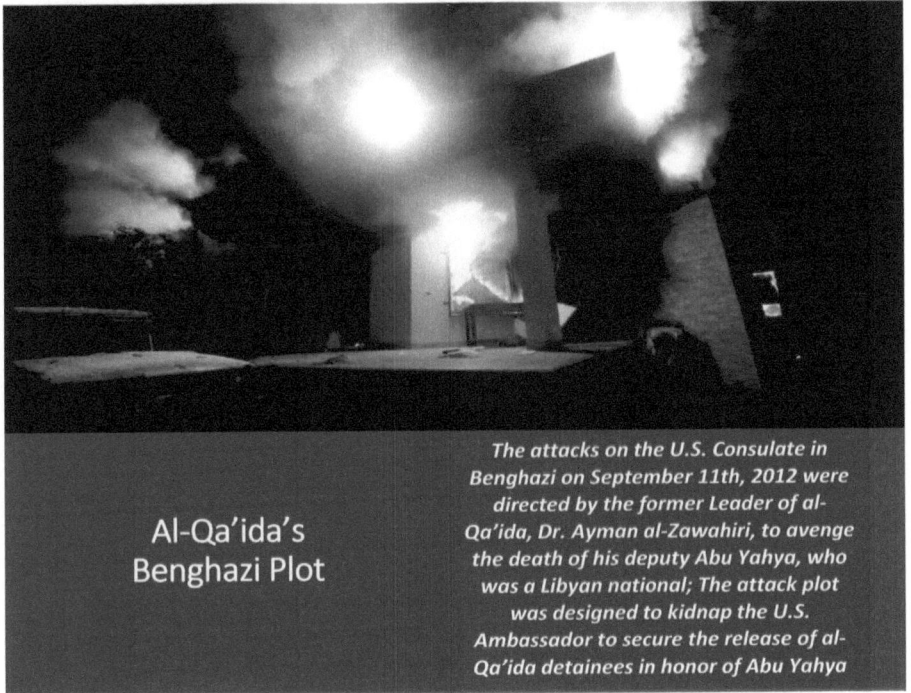

Al-Qa'ida's Benghazi Plot

The attacks on the U.S. Consulate in Benghazi on September 11th, 2012 were directed by the former Leader of al-Qa'ida, Dr. Ayman al-Zawahiri, to avenge the death of his deputy Abu Yahya, who was a Libyan national; The attack plot was designed to kidnap the U.S. Ambassador to secure the release of al-Qa'ida detainees in honor of Abu Yahya

As with most stories regarding al-Qa'ida, since the September 11th, 2001 attacks, this story begins in the border region of Pakistan and Afghanistan, a connection almost always downplayed and often completely ignored when discussing the U.S. Consulate and CIA Annex attacks in Benghazi. This book will set the record straight, name names, and properly place responsibility on those that planned and carried out the Benghazi attacks—Al-Qa'ida's Senior Leadership (AQSL).

The first place to start is at the top. In 2012, Dr. Ayman al-Zawahiri (known hereafter as Zawahiri) was the Leader of al-Qa'ida. He ran the terrorist group primarily from the Pakistan side of the border and held the position as Leader from 2011 to 2022. It was Zawahiri who directed the attack, but he strategically also included the heads of all the affiliates of al-Qa'ida in the operational planning efforts.

Al-Qa'ida's Benghazi Attack Planning

What sets the Benghazi attacks apart from most attacks, showing the importance Zawahiri placed on the attack plotting, was that every one of al-Qa'ida's direct affiliates was involved in the attacks: AQIM, al-Qa'ida in the Arabian Peninsula (AQAP), al-Qa'ida in Iraq (AQI), and then newly formed al-Qa'ida in Egypt (AQE). See the following graphic, showcasing the global al-Qa'ida affiliates involved in planning, financing, facilitating, and carrying out the Benghazi attacks.

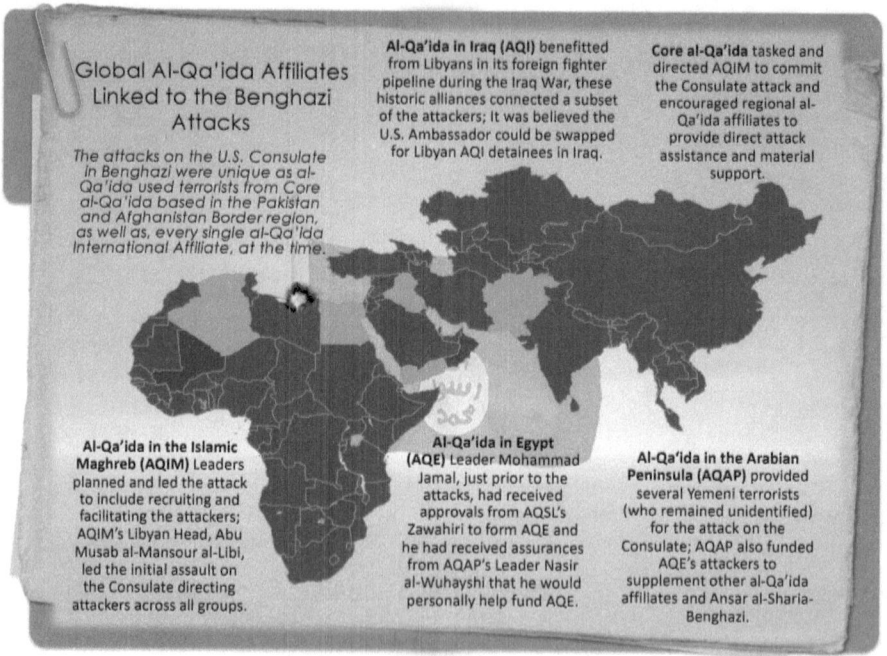

Zawahiri planned the attack against the U.S. diplomatic presence in Benghazi after the demise of his deputy, Abu Yahya al-Libi. Abu Yahya became al-Qa'ida's second-in-command after Zawahiri was elevated to Leader following the long-awaited death of al-Qa'ida founder Osama Bin Laden. Before his promotion to Deputy, Abu Yahya had been the Operational Leader for al-Qa'ida in the Pakistan, Afghanistan, and Iranian border regions. He started as a former leader in the Libyan Islamic Fighting Group (LIFG). A CIA target is placed on the head of anyone holding any position in AQSL, and on June 4th, 2012, Abu Yahya was killed in Pakistan. The White House responded to his death by noting that it "dealt a heavy blow to al-Qaeda's operations" as he played a critical role in the group's planning against the West and maintained contacts with al-Qa'ida affiliates across the globe, including North Africa.

The death of Abu Yahya was not a one-off, it was essentially the final straw for Zawahiri, who in just a year had lost his Leader and long-time confidant, Bin Laden, as well as his first two deputies. In addition to Abu Yahya, Zawahiri's first deputy, another senior Libyan named Attiyah Abd al-Rahman (with real name Jamal Ibrahim Ishtiwi al-Misrati and also known by Abu 'Abd al-Rahman and 'Atiyyatullah) was killed in a U.S. drone strike in Pakistan on August 22nd, 2011, less than ten months before Abu Yahya was killed. Before being promoted to Zawahiri's Deputy, Attiyah was one of the leading public faces of al-Qa'ida. Attiyah had also been a senior leader in LIFG alongside Abu Yahya.

It is important to note that since the attack planning started in June 2012, there was no connection to the film *Innocence of Muslims*, which sparked off violent anti-U.S. protests in several locations in September 2012. The video had not gone viral before the attacks in Benghazi. It was, however, the blaming of the video by the U.S. Government that caused the video to go viral. The video had just over 100,000 views early on September 12th after the completion of all six attacks on the U.S. Consulate and CIA Annex in Benghazi. An online video on the Internet must have a total of 5 million views over 3 to 5 days to be deemed viral. After U.S. Officials continued to blame the video for the attacks and even some attackers seized on the opportunity to piggyback on this narrative, the video still was only at 1.9 million views at the end of the day on September 13th, 2012.

Back to the start of planning in June 2012. Zawahiri tasked Abdel Malek Droukdal, the Leader of al-Qa'ida's affiliate in Algeria, AQIM, to develop an appropriate attack plan against the U.S. in Libya. The key communications from Zawahiri regarding the attack ran through his brother Mohammad al-Zawahiri who had visited and maintained ongoing and direct contact with AQIM's Droukdal starting after Mohammad's release from prison in Egypt (first in March 2011 and then again in March 2012). Mohammad also maintained contact with Egyptian Mohammad Jamal, the Leader of newly formed AQE. Jamal had a long association with al-Qa'ida leader Zawahiri as they both served in the Egyptian Islamic Jihad (EIJ). Jamal also served in Yemen on behalf of EIJ and maintained extensive ties to AQAP during the Benghazi attacks. AQAP had provided funding for the U.S. Consulate attack.

Droukdal, like Zawahiri had a personal connection to Abu Yahya, as they had been close friends for approximately twenty years after fighting together in Afghanistan. Therefore, the initial plan designed by AQIM was to capture the Ambassador, as Stevens was not only the symbol of America in Libya, but he was also a symbol of American influence, primarily to Eastern Libya, due to his highly visible role in NATO's intervention in the Libyan revolution. AQIM was a logical choice for Zawahiri's plot as the group was the al-Qa'ida experts in kidnapping operations.

For AQIM, kidnappings usually generated a great deal of the group's funding, this tactic also yielded propaganda content—which led to more members, followers, sympathizers, and even more money. While al-Qa'ida knew the U.S. would likely not pay a ransom, they had found a commodity just as valuable—prisoners. The Arab Spring had given terrorists hope that they may no longer be locked up in prisons for life, and as al-Qa'ida members started to get out of prison, they became almost obsessed with not leaving their allies behind who were still imprisoned. Al-Qa'ida felt that freeing more of its terrorist allies would be the right way to honor Abu Yahya.

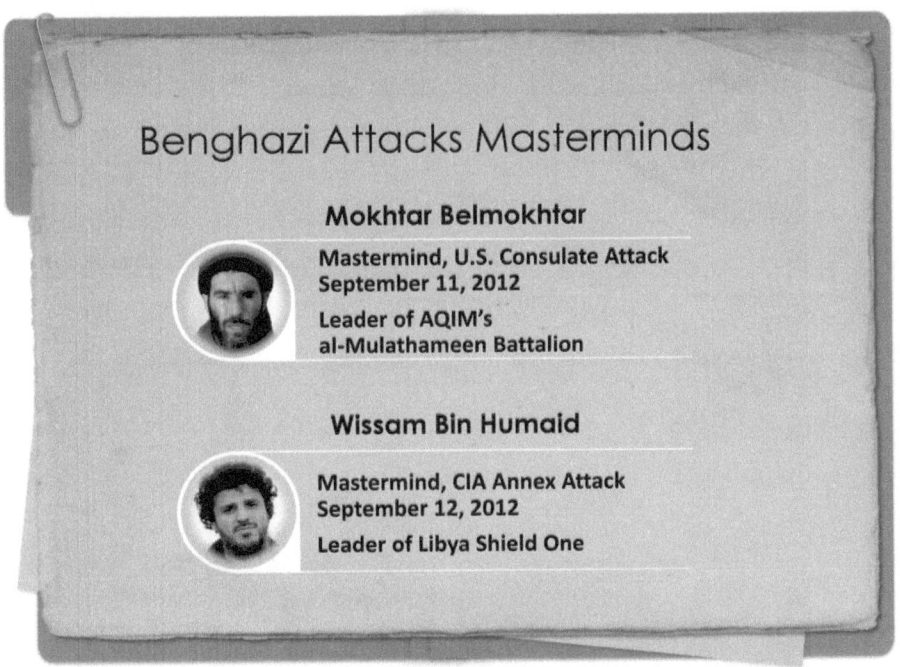

After being tasked by Zawahiri in early June, Droukdal had the option to task one of two key operational deputies, referred to as AQIM Battalion Commanders, to resource a kidnapping plan. One of the operational deputies was Abdelhamid Abu Zaid (also known as Mohamed Ghadir), and the other was Mokhtar Belmokhtar (known hereafter as MBM). Droukdal leaned in early after the Libyan revolution and sent a number of operational commanders to the region to establish key relationships. Abu Zaid and MBM were at the forefront of these efforts connecting with senior terrorists, militia leaders, and Libyan Government figures promptly after the death of Muammar Gaddafi on October 20th, 2011.

MBM was a close associate of Zawahiri as he preferred to report directly to Zawahiri instead of Droukdal; and as luck would have it for MBM, he was in Libya, specifically Benghazi, when the death of Abu Yahya was announced and eulogized by AQSL. So MBM got the charge to mastermind the attack on the U.S. Consulate in Benghazi and later would employ key individuals in his al-Mulathameen Battalion, and within greater AQIM to plan, to recruit fighters, and to carry out the attack.

We refer to MBM as the Mastermind of the attacks on our U.S. Consulate[i] in Benghazi on September 11th, 2012, as there is no evidence that al-Qa'ida was involved in the attacks on us at the CIA Annex[ii] (a graphic of the two locations follows). Therefore, the CIA Annex attacks on September 12th, 2012 had a separate Mastermind, Wissam bin Humaid, who was the Leader of Libya Shield One. The Libya Shield was a conglomerate of militias funded by the Government of Libya in lieu of the formation of a real national military.

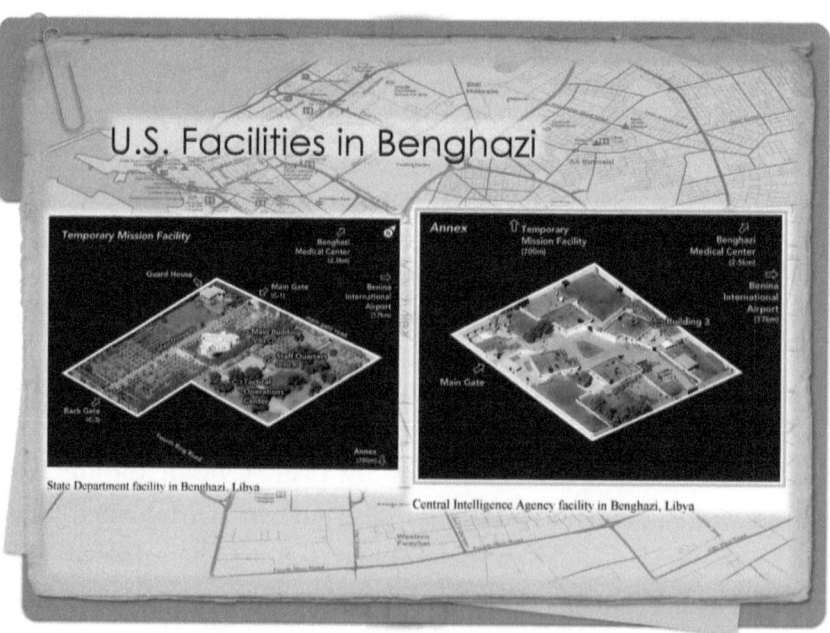

On June 7th, 2012, MBM was in Benghazi to attend the first annual conference hosted by Ansar al-Sharia-Benghazi (AAS-B). This so-called "conference" was hundreds of armed groups holding a demonstration in Benghazi in support of Sharia law in Libya. More than 1,000 Islamists attended from various militia and terrorist groups in the region. AAS-B had a military parade in the streets of Benghazi, which alarmed our CIA personnel in the city due to its size and the vast number of heavy arms on display. MBM traveled to Benghazi for this military parade after a personal invite from Wissam Bin Humaid to be his esteemed guest.

Here we had the two future Masterminds showing off their guns and acting like rulers over a city that was not theirs and never would be, thanks to future counterterrorism operations by General Khalifa Haftar and his Libyan National Army (LNA). Unlike AQIM, who planned the U.S. Consulate in Benghazi kidnapping plot in advance and likely made Wissam privy to those plans, Wissam took advantage of the crisis that night to strike a blow at the CIA. The sole purpose of the attack was to get Americans out of Benghazi for good. He had put up with American influence long enough in Benghazi, an influence he believed harmed getting political candidates elected that had been proponents of militias, like his own, retaining power.

One may wonder about the "Mastermind" of the Benghazi attacks arrested and charged by the Federal Bureau of Investigation (FBI), Ahmed Abu Khatallah (known hereafter as Khatallah). While Khatallah was a known militant in Benghazi and arrived at the Consulate after being notified by an associate that an attack was underway, it is a complete fallacy that Khatallah was the Mastermind. Khatallah was so far out of the loop, so to speak, that he was sipping tea at his house, oblivious, when the attacks started on September 11th, 2012. As such, we will not perpetuate misleading claims about his role.

While al-Qa'ida's involvement is downplayed in the attacks, it has been well-known that AAS-B was the primary local Benghazi-based group to provide attackers. The group, an umbrella organization of multiple groups, boasted about 5,000 members by the end of 2012. By mid-June, it had already proven they could overrun a Consulate when a group of 20 armed AAS-B members overran the Tunisian Consulate in Benghazi. They ended the attack by hoisting their black flag over the Consulate. AAS-B was founded in February 2012, so it was newly formed when MBM pulled the group into AQSL's and AQIM's plans related to Benghazi.

Although the group had not yet proven itself a terrorist organization, it was a chosen partner due to its leader, Mohammad al-Zahawi, and his historical relationship with al-Qa'ida Founder Osama Bin Laden. As Zawahiri directed this plot, every level of attack planning required the utmost trust that only came with historic, vetted relationships. The U.S. Consulate in Benghazi was an easier target for al-Qa'ida than the U.S. Embassy in Tripoli, not only because the Consulate compound lacked appropriate security but because the group had insiders working in the Consulate. Further, al-Qa'ida had many affiliates to pull from in Benghazi, first due to local terrorist support for other al-Qa'ida campaigns, primarily in Algeria, Iraq, and Afghanistan, but also as al-Qa'ida had been establishing a base in Benghazi since late 2011. The group had already formed local relationships, set up training camps, and had an operational infrastructure to draw from in the city. One major unknown, though, was when Ambassador Stevens would be in Benghazi.

Chapter 2

Our Libyan Arrivals

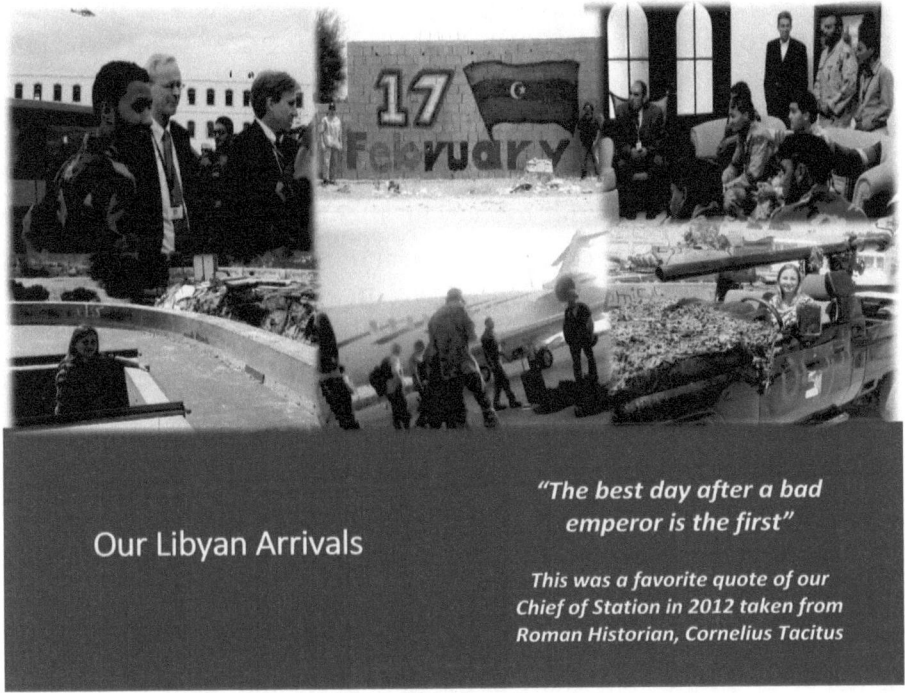

Our Libyan Arrivals

"The best day after a bad emperor is the first"

This was a favorite quote of our Chief of Station in 2012 taken from Roman Historian, Cornelius Tacitus

We both flew into Libya in 2012, but to two very different cities that looked and felt worlds apart. Boon was planning to go to a country in the Middle East and then was asked to change his deployment to Benghazi, Libya. I then ended up changing my second deployment to Libya and joined him in Benghazi in early fall 2012. With Libya, we knew we were walking into not just a post-revolutionary country but a country without an established government.

Boon's Point of View

It was just before my birthday in July of 2012. I was suddenly awakened by the sound of the pilot talking over the intercom, advising the flight attendants to prepare the cabin for landing. I slid the plastic cover-up over the window to see the Mediterranean ocean and a vast desert. I had slept most of the plane ride, which I usually did since we were expected to hit the ground running. "Hit the ground running" generally means working at least a twelve-hour day and hoping for a shower at some point. After exiting the aircraft and being hit with the hot, humid wave of air, I made my way into the chaotic sea of humanity, rushing across the tarmac to baggage claim. This type of third-world country's unorganized chaos was nothing new to me, as I was already in my 8th year with the Agency and had thousands of days of deployments. I easily blended in with the locales as I waited for my ride to show up after retrieving my bag from the revolving circus that served as baggage claim.

Scanning the entrance at Benghazi's Benina International Airport, I immediately picked out my GRS counterpart, who had come to take me back to the CIA Annex. Without saying a word, I started to move toward him. When he saw me, he simply turned around and started to walk out the airport entrance toward our vehicle as I loosely followed in trace. Once inside the vehicle, he smiled and extended his hand, saying, "Welcome to Benghazi, brother. I am Rone. Heard good things about you from a few guys back at base."

A few of my teammates were already in the country. In the world of warzones, I have seen many of the same faces and many of the same smiles. Leaving the airport, the landscape of small desert dunes and farmland immediately made me think of Tatooine from Star Wars. It was quite fitting that the movie was filmed just across the border in Tunisia. This brought me back to a photo Sarah had sent me months earlier from a place called Villa Silene that Glen "Bub" Doherty had taken of her over in western Libya that we both joked was also from Tatooine. Apart from a few checkpoints, the ride to CIA's Annex in the city, which we referred to as "Base," was uneventful, which was always preferred.

Sarah's Point of View

In January 2012, unlike Boon, I had about a week's notice before my first trip to North Africa, starting in Libya. The short notice led to several administrative issues I had to overcome. Acquiring a visa on such a quick timeline proved difficult, resulting in having to fly out with a letter and a promise by the Libyan Embassy that I would be issued a visa upon arrival. I arrived at Tripoli International Airport very tired. I remember it being very quiet, dark, and empty. There was no sort of passport control or security of any kind. This would later pose a problem for me when I left the country as I did not have a stamp in my passport upon entering, and they did not want to allow me to leave.

I saw a few Zintani militiamen as I roamed the airport, realizing that anyone could easily enter the country and that my visa letter was likely useless. The militiamen had zero interest in me being there. Luckily, I finally contacted a Senior Libyan Official who welcomed me to Libya by telling me his whole history in protocol related to foreign dignitaries. I could tell he was the guy who could grease the skids and get things done even in a volatile Libya, so I knew we would meet again. He guided me out of the airport, and then I was in Tripoli, climbing into a vehicle in the parking lot to also head to the CIA Annex.

The Senior CIA Officer in Tripoli was very familiar to me as he had been Head of the Pakistan and Afghanistan Department (PAD) in the Agency's Counterterrorism Center (CTC). PAD was where I kicked off my CIA career. Still, we had never worked closely until I landed in Libya. We hit it off immediately as he had a multitude of good ideas to help the Libyans in their transition, and we were both ready to go after the terrorists in country and those flooding into the country.

I was asked to come out to Libya to help jump-start capture operations against terrorists, primarily MBM. The Chief was optimistic that we would make it happen quickly, but we first had to get through the big hurdle that was CIA Headquarters (HQS) who clearly told me they would not give us any permissions to go after terrorists in Libya.[iii]

Also, we needed to also get the Libyans to understand the current terrorist threat and be tough on terrorism from the start so extremists would not view it as a safe haven. That Spring became a masterclass in risk, innovation, governmental affairs, and negotiations. I would go back to Libya in August, but to our Annex in Benghazi which was none of

Our Libyan Arrivals

those things.

Tripoli felt like being in southern Europe; Benghazi felt more like a city in a Middle Eastern warzone. Where I would see trucks with Texas license plates riding around Tripoli while sitting at a coffee shop, Benghazi had black flags, signifying extremist takeovers all around town. I had to be very comfortable in both cities with gun trucks and large weapons. In Benghazi, there was almost a militia attire for men, usually consisting of a mismatch of military fatigues and sometimes skinny-cut t-shirts. In Tripoli, more men dressed in casual clothes like slacks and polos unless at a militia checkpoint.

September 11th, 2012, should have been a typical day. I flew off to Europe, then, just after midnight, I called back to Libya to check in. Boon answered from the roof, saying he was busy but safe and couldn't talk at the moment. I said goodnight, not realizing at the time what was going on. My alarm went off at 0500 in Europe, which was 0600 in Benghazi. I had a million messages from my mom, who was normally a one-text person at most. She knew I was in Europe, but she asked about everyone in Benghazi. I grabbed the remote and started flipping through channels on the hotel's television. I then went through every number in my phone, resulting in unanswered calls or dead phones. I finally reached a good friend, a member of the GRS team. He told me they had been attacked and that Ambassador Stevens and Sean Smith were dead, as well as two CIA officers. I was in complete shock.

He passed the phone over to Boon, who repeated the information. I pressed him to tell me who we lost. He finally gave in and said, Rone and Bub. So, I started arguing with him, saying, "can you repeat that? You are saying the wrong name; Bub wasn't in Benghazi". I did not even know Bub was back in Libya. When we'd last worked together, he had gone home to San Diego, where he had been hit by a car while riding his bike, and I thought he was still home recovering. I teared up over Rone as he had been our defacto leader in Benghazi and was the person we all trusted to keep us safe while there. I kept pushing back on Bub's name, mostly out of shock and because I did not want to believe either of them were killed.

Finally, Boon told me a team had come from Tripoli, and then I realized what I was hearing was a fact. The worst part was I had to get ready and go to work when the only place I wanted to be was back with everyone in Libya. I arrived for scheduled meetings at 0845 and met with our Deputy Chief of Station (DCOS), aka "Frat boy" from Tripoli, and an analyst who had been temporarily helping out in Tripoli. Neither of the two understood what had transpired at both the Consulate and CIA Annexes. I told the DCOS that we lost Rone and Bub and that he needed to go to the airport to head back to Tripoli. The analyst and I could handle these European meetings, and I wanted to get a flight back immediately. The DCOS said no and would stay in the country as he had "dinner plans."

After the three of us left work for the day, we first had to go watch shopping. No exaggeration, the DCOS was looking for a new watch, and then he brought us to his miserable dinner with some young CIA officer he used to work with—who was clearly, pathetically in love with him. Years later, I ran into the DCOS in the CIA cafeteria when I was working for Congress. When he saw me, the first thing he said, "you know that I know nothing about the attacks," in an effort to ensure we weren't subpoenaing him to testify in Congress. I responded, "oh, I know" (as I ended my phrase by saying "frat boy" in my head). To his credit, though, he apologized for his actions that day in Europe, which I appreciated. Still, it was less about how he made me feel that day and more that he showed no care that we lost Rone and Bub—and remember, Bub was one of the officers protecting him in Tripoli.

The Ambassador

Ambassador Stevens returned to Tripoli in May 2012 after serving in Benghazi during the 2011 Libyan revolution. He first intended to travel to Benghazi in August 2012, a trip we were preparing for. Due to the increasing violence in the east, the trip was canceled. It was unknown if al-Qa'ida was aware of this trip and had to pivot, but it is more likely the trip was canceled before details on it were leaked to local elements in Benghazi.

The next opportunity for Stevens to travel was in mid-September 2012, when the Principal Officer in Benghazi was going on leave. It would leave an opening for someone to fill in from the U.S. Embassy in Tripoli.

Unfortunately, the security situation in Benghazi had significantly worsened since August. Despite the three largest militias indicating they would pull most of their security support to Americans in Benghazi, Stevens was still eager to come back. As background, on April 2nd, 2011, Stevens arrived in Benghazi via a Greek cargo ship during the throes of civil war. He arrived at the base of operations for the rebel-led National Transitional Council (TNC) to re-establish a U.S. presence in Libya as the U.S. Special Envoy. The connections made during this time meant Stevens knew most of major players we will mention involved in the attacks, based in Benghazi.

Besides nostalgia, Stevens was also on an administrative time crunch related to Secretary of State Hillary Clinton as she was tentatively planning to travel to Benghazi in October 2012. This would be her first trip back to Libya since the death of Gaddafi, and Stevens wanted to prepare locally for her visit. This trip by the Secretary was important to Stevens to help secure funding support for a new joint CIA/State Department facility in Benghazi. It was vital to Stevens that Benghazi felt equal to Tripoli in the support provided by the U.S. Government, as Gaddafi had long sidelined the city.

On September 10th, 2012, after settling in at the Consulate, the Ambassador traveled to the CIA Annex for a counterterrorism briefing. Stevens, who arrived in Benghazi in April 2011, was quite familiar with our Annex as he lived there after the Tibesti Hotel was bombed on June 1st, 2011, where he had maintained his base of operations. Stevens resided at the Annex for three weeks until he relocated to the new U.S. Consulate in Benghazi compound for the first time.

The Ambassador showed some concern as we briefed him on Benghazi's worsening security situation and threat environment. He was not adequately briefed on the current counterterrorism threats before his arrival. The fact that an Ambassador was surprised and unaware of how bad the second largest city in his country had gotten should be chilling to anyone charged with collecting and disseminating vital terrorism and security-related information.

Most of us CIA Officers were at the briefing to the Ambassador, except our COB Bob and his security detail. Bob expressed to us that he was annoyed that the Ambassador was in town and did not want to have to interact with him. Bob told the CIA Staff at the time that the leader of 17 February Martyrs Brigade (known hereafter as "17 February"), Fawzi bu Khatif, was leaving the country early in the morning of September 11th. (Later realized to be untrue, as Fawzi was in Benghazi when the attacks started late on the 11th.) Bob claimed it would be his only opportunity to see Fawzi before he left, so that meeting was more important than seeing the Ambassador.

Bob met Fawzi alone, as he did not invite his security detail inside the meeting. While Fawzi likely knew of the impending attack the following evening, he did not share any threat reporting with Bob.

Our best assessment is that al-Qa'ida had at least five days' notice before September 11th that Ambassador Stevens would soon arrive in Benghazi. On September 6th, 2012, the State Department sent an official request to increase local police security support to the U.S. Consulate from September 10th to September 15th to the Libyan Ministry of Foreign Affairs and International Cooperation, Office of Public Protocol in Benghazi.

On the same day, we received information at the CIA Annex that some senior terrorist leaders had traveled to Benghazi for a high-level meeting. At the time, the reason remained unknown. In hindsight, al-Qa'ida might have finalized the attack plans on the 6th or 7th. While awaiting his arrival, al-Qa'ida and its allies had already collected information on the comings and goings of the U.S. Consulate in Benghazi staff and the security deficiencies at the Consulate through their allies in 17 February. The attackers even had passports printed in fake names to flee after the attacks, so they were well prepared to carry out the attack on short fuse notice.

Of note, the police service, called the Supreme Security Council (SSC), provided by the Libyan Government's Ministry of Interior (MOI) to secure the vicinity outside of the Consulate compound, was complicit in the attacks. The SSC was involved in the early morning casing incident outside the Consulate. Then, they provided cover for the attackers as they coalesced outside the Consulate before the attack. SSC departed just prior to the attacks, not alerting the State Department of the impending assault.

Even more concerning was that a senior SSC Official (the CIA redacted his name) reportedly received advance notice of the attack. At the time, he attempted to contact three separate individuals at approximately 1700 local time in Benghazi on September 11th, 2012. This would have provided ample warning to evacuate personnel from the U.S. Consulate, as the attacks did not occur until 2142 local time.

First, he contacted the head of the Libyan Intelligence Service (LIS); however, he was on an official Libya Government visit to Rome, Italy, and was unreachable. Second, he attempted to reach out to the LIS Chief of Staff, also in Rome. Lastly, he tried to contact our Chief of Base (COB) referred to as "Bob" whose phone was off. Bob had recently been annoyed with the SSC Official as he requested financial assistance to help take care of a sick family member. As such, for at least three weeks prior to the attacks, Bob had ignored him and provided no alternate way for the Official to contact the CIA with threat reporting.

On September 15th, 2012, Bob and another CIA Case Officer spoke with this Official, who reported his failed contact attempts. Bob became enraged and tried to choke the Official and needed to be pulled off him. To cover up this information, when Bob wrote up his intelligence report, he only claimed that the Official failed to pass the notice of the threat to the LIS. The report also stated that this Official made "no" attempts to pass the information on to U.S. personnel. As Bob had remained in Libya collecting information after the attacks, it is unclear what else may have been excluded from CIA operational channels and CIA disseminated intelligence reports.

CIA's Chief of Base

I met Bob briefly in February 2012, but the first interaction I can remember with him in Benghazi was a disagreement over the LIS. As background, when I was in Tripoli, I met with the Head of LIS when the organization was formed in spring 2012. About a month in, I asked him how many LIS officers were in Benghazi; he said several hundred, which was good to hear as the organization was newly formed. However, when I arrived in Benghazi in August 2012, I asked Bob how many LIS Officers were in Benghazi, and he said two. Yes, two, I said that makes no sense. I then asked where their headquarters was, and he said they did not have an office and only had a room at the Tibesti Hotel in Benghazi.

So, I go to the hotel office with the Bob and the newly arrived DCOB. I walked into the room, and kid you not—there was a rotary phone sitting on the desk with no cord coming out of the back and a laptop not plugged in as well—that's it. So, I look directly at Bob and say, "this clearly isn't their office." The DCOB and I asked, and LIS noted they had 500 officers in Benghazi. I then asked if we could meet at their Benghazi headquarters, and they said, "of course, Bob had never asked before." For six months, Bob did not know where the office was of his key Libyan partner.

Fast forwarding to September 11th, from the start, Bob was delusional during the attacks. For example, he forgot to report that a terrorist attack was occurring in Benghazi to the COS in Tripoli and the CIA's Headquarters. The most egregious issue, though, happened immediately after the attacks kicked off. Most people think it was the standdown; frankly, Bob was never going to let us go. He even lied to keep us there, we finally left, but the "lie" is what we need to share.

The actual issue was that in Bob's first call to the 17 February leader, Fawzi bu Khatif, Fawzi immediately told Bob that he had no resources available to send to help support us in defense of the U.S. Consulate. 17 February had a least 3,000 members and resources that rivaled a modern army's resources, so it wasn't can't help, it was won't help. Bob withheld the information from us during the entire duration of all six attacks. He not only withheld it but also lied to us and said to stay at the Annex and wait for 17 February to respond. He told us multiple times that 17 February was sending a response. His withholding caused a deadly delay and threatened the safety and security of our response team as we were cautious when interacting with actual terrorists because we were concerned about a blue-on-blue type of incident with 17 February.

One last item: I had gone over to the U.S. Consulate a couple of days before the attacks occurred with two of the GRS Officers. The purpose of the meeting was the fact that the State Department agents did not trust the local guards on their compound grounds. They asked us if we could perform technical collection against the Consulate to see if we could connect any of the guards to terrorist groups like al-Qa'ida. On the morning of September 11th, GRS went back to perform another technical collection at the Consulate.

After GRS was freed from their involuntary detainment in Germany, one of the GRS Officers reached out and asked if anything positive came of their collection. I had no idea what he was talking about, and he said he gave a flash drive to Bob. He told Bob it was vital as it could potentially identify attackers if they were casing or pre-staging in the area earlier in the day. Timing-wise, this is now a week after the attacks. Bob had never provided this flash drive to anyone in the CIA at Tripoli Station. I immediately go to Bob and ask him to turn over the drive. He notes that it was in his room, but he had no idea what it was.

When our Base in Khowst was attacked, the female COB was blamed all over the media for the loss of CIA life on the compound—there was even a week-long lessons-learned course created from the events. In contrast, Bob was given the Distinguished Intelligence Medal (DIM)—CIA's highest honor.

Chapter 3

The Rising of the Black Flags

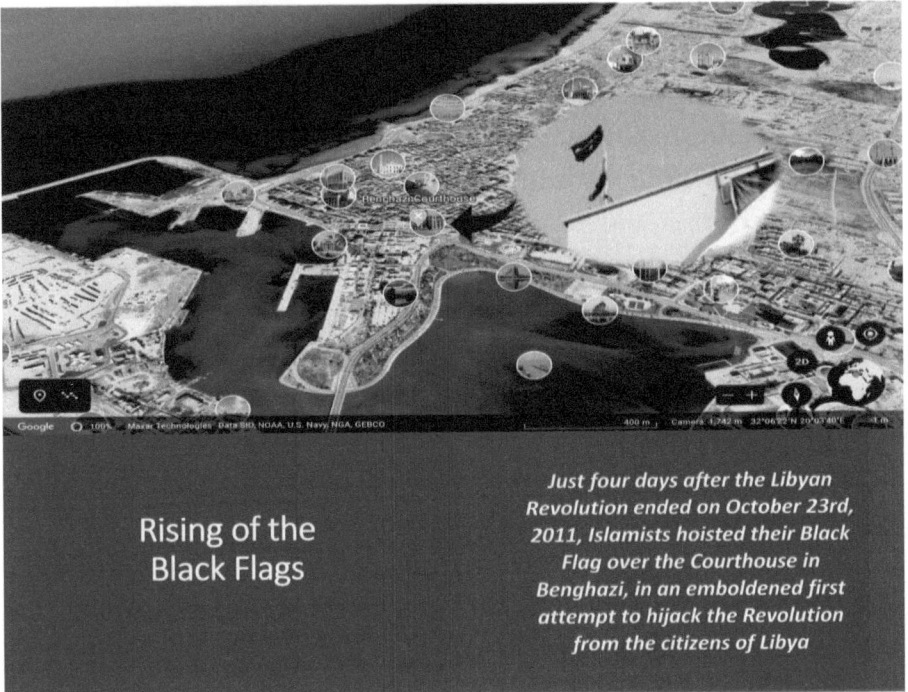

Rising of the Black Flags

Just four days after the Libyan Revolution ended on October 23rd, 2011, Islamists hoisted their Black Flag over the Courthouse in Benghazi, in an emboldened first attempt to hijack the Revolution from the citizens of Libya

In Benghazi, a moment signified the city's shift to extremism in late October 2011 when a black militant flag started flying over the Courthouse in Benghazi. Writer Sherif Elhelwa, who wrote for the Vice on October 27th, 2011, explained the significance of the Courthouse best, "It was here at the courthouse in Benghazi where the first spark of the Libyan revolution ignited. It's the symbolic seat of the revolution; post-Gaddafi Libya's equivalent of Egypt's Tahrir Square. And it was here, in the tumultuous months of civil war, that the ragtag rebel forces established their provisional government and primitive, yet effective, media center from which to tell foreign journalists about their "fight for freedom."

This was the day that extremist revolutionaries in Benghazi made it known to the public the threat they posed post revolution. They expressed that they wanted an Islamic State and Sharia law to rule the land. The

Islamists did not intend to give up their militias and weapons nor join official government bodies like the army or the police. The plan was not for a Benghazi that the West supported and defended. The plan was for a Benghazi and a greater Libya to be ruled by the militia leaders—but not heroes from the revolution—just those who were Islamists. Those who spoke with the loudest voices, even if their voices were at odds with what the majority of Libyans actually wanted.

Since coming to power after a successful coup d'état in 1960, Gaddafi understood the threat terrorism could pose to his rule. He handled the threat of terrorists and Islamist-affiliated groups like LIFG, Martyrs Group, EIJ, al-Qa'ida, and other dissidents with a heavy hand.

To balance this, one must understand that how Gaddafi ruled was the main cause of the terrorist insurgency in his country. Gaddafi himself promoted terrorism and benefitted from using terrorist groups and terrorist acts when he thought it would advance his goals, particularly in negotiations with other governments and in his attempt to rid himself of other governmental rivals through assassinations.

Well before the late 1980s, starting just after he came to power, Gaddafi hosted terrorist training camps and provided training and weapons for several groups on the U.S. List of Designated Foreign Terrorist Organizations. These groups included the Basque separatist group ETA, the Irish Republican Army, the Japanese Red Army, Italy's Red Brigades, the Palestine Liberation Organization (PLO), the Moro Muslim Guerrillas in the Philippines, and Sierra Leone's Revolutionary United Front. The most well-known terrorist attack Gaddafi was involved in was the bombing of Pan Am Flight 103 over Lockerbie, Scotland, just before the Christmas holiday on December 21th, 1988, which killed 270.

According to the CIA, the bombing was widely regarded as an assault on a symbol of the U.S., with 189 of the victims being Americans, and stood as the deadliest terrorist attack on American civilians until September 11th, 2001. One of those killed was our CIA acting Deputy Chief of Station (DCOS), Matthew ("Matt") Kevin Gannon, posted at the time in a temporary duty (TDY) position in the Middle East. Matt's name was revealed for the first time just months before the

Benghazi attacks, on May 21st, 2012, during the CIA's Annual Memorial Ceremony. CIA's former Director David Petraeus provided the following comments on Matt's service and sacrifice: "His deft tradecraft skills, superb language ability, and strong interest in Middle Eastern cultures would have put him on course to be a leading officer in the war against al-Qa'ida and its affiliates." Matt's star is on the same wall where we watched both Rone and Bub's stars etched. As stated best by the CIA, his star joined an honored constellation of souls, fallen officers whose courage, integrity, and devotion to the safety and security of our nation will never be forgotten. "In them," said Petraeus, "we saw what is best and most admirable about our Agency and, indeed, our country."

In contrast, the extremist threat in Benghazi was known on the city streets the same week that NATO officially declared operations ceased on October 31st, 2011. Islamists who were patient during an unwanted NATO intervention would not wait nor allow the West to affect the power and leadership positions they believed now belonged to them in Benghazi. The revolution may have started to get rid of Gaddafi, but for the Islamists, it became more focused on what they wanted for themselves. They hijacked the democratic ideals of everyday Libyans, a people who had risen against Gaddafi to build a better future for their country.

When the black flag went up over the courthouse, everyone in Benghazi knew what it meant. It is important to understand that the revolutionaries did not want the West to have influence, especially as they finally had a powerful voice in things not afforded them under Gaddafi. Most critically, they built their powerful armies that did not just help win a war but now helped rule a city, their city, Benghazi.

Sadly, to win against Gaddafi, all the enemies of Gaddafi in the country had to be friends, but then after the revolution, their ideologies understandably pushed them back apart. While a fringe idea at the time in Libya, terrorism still got a foothold as the revolutionaries needed the skills, tradecraft, resources, and personnel from Libyans who had been trained at terrorist camps and involved in armed conflict in foreign wars. Gaddafi had not allowed citizens to own weapons. Therefore, those most proficient in the use of arms, tanks, and missile systems were those that had been arrested and jailed under Gaddafi for their roles in supporting terrorist groups like EIJ, al-Qa'ida and the LIFG fighting outside of Libya like in Algeria, Afghanistan, Yemen, Iraq and Syria.

One of the first things we do in GRS when traveling to a new area of operation, which Benghazi was for me in July 2012, is what we call area "familiarization." This familiarization not only includes learning the roads and traffic patterns but also the atmospheric conditions of the city; the sentiments of the locals in the different areas and neighborhoods; the varied political and religious views; tensions between different groups, militias, and/or tribes; potential safe havens or allies we could affect; and avenues of ingress or egress into certain areas of interest.

The black flag was just the start. There were hundreds of attacks in Benghazi from the end of October 2011 into the lead-up to the first attack on September 11th, 2012. The flag symbolizes the start of a new war. Attacks in the city gradually became more and more common as the weeks passed that summer of 2012. One day during an area familiarization, our team noticed the black flag flying in a more affluent and normally pro-western neighborhood occupied by many Libyan government officials and their families. This was alarming as in GRS; we already had to rely on a very uneven ally in 17 February, which was known to support Islamists. It became more concerning as we had to worry about Islamist influences within the actual Government of Libya. A decade later, this problem is still a major concern and it puts our current diplomats in the same danger we were in—as they may not clearly know or understand who their actual enemy is.

It is hard to bash the Libyans when our government, including our former President Barack Obama, noted that failing to prepare for the aftermath of the ousting of Libya's Gaddafi was his worst presidential mistake. It is easy to make mistakes when patriotism is involved. All those who fought were heroes in the immediate aftermath of the revolution. These heroes and militiamen had earned trust and respect for their sacrifices. After the American revolution, both sides came together to be Americans; but this was not the case for the Libyan revolution. Unfortunately, while many good people were trying to establish a legitimate government to support democracy in Libya, they failed to recognize Islamists' actual threat to this process. The country did not come together to form a government that would be successful.

It was not just Libya not understanding this issue, but the U.S. failed to understand the shifts. We supported a revolution and allowed terrorists from al-Qa'ida and EIJ to fight shoulder-to-shoulder alongside our interests; however, we had no endgame strategy of how to sideline the terrorists and their control post revolution. How do we get them to disarm when they have broader goals for the future of Libya? How to stop the influx of more like-minded terrorists when a government wasn't established? When a police force did not yet have its gear? When there were more militias than actual government military personnel? When an intelligence service was starting from the ground up?

It was easy then for terrorists to move in and out of Libya with their new government institutions, especially terrorists like MBM, who the U.S. and Interpol wanted. The institution I got to watch firsthand be formed was the LIS, which again was the Libyan Government's counterpart to the CIA. LIS was headed by a gentleman who had been a professor in the State of Georgia before the fall of Gaddafi. From interactions with him, he was the kind of man one would want as President. He thought strategically, loved his country, believed in his people, and was calm, steady, and measured. A member of Amnesty International, he made it clear that he wanted the country to handle terrorists and extremists in an opposite way of how Gaddafi had, which was a good thing. Still, with so many other issues the intelligence agency faced, there wasn't time to take a 20-year approach to a terrorist problem that was here and now.

Upon its formation in the Spring of 2012, the LIS viewed two major problems in the country. The first was a migrant issue, as Libya was used as a thoroughfare for other African migrants on their way to illegal immigration in Europe. The second major issue was the worry over the number of former Gaddafi supporters. Many of these Gaddafi supporters were operating underground, and the fear was whether they would be successful in building opposition to attack the newly formed democracy. We had a personal interaction with one of these underground Gaddafi networks, it was members of his former military intelligence who came to our aid and helped us evacuate to the airport the morning of September 12th, 2012 after the lethal mortar strikes on the CIA Annex.

The revolutionaries were supplied salaries by the government for their service, which naturally also meant the terrorists who fought in the revolution were receiving salaries, as well. Libya had long operated in a socialist manner, so it was expected that the government would provide for these fighters. The government had also hoped that this would help the fighters leave their alliances within the many militias across the government and join in official government organizations like the Libyan Army (which really never took off in the official Libyan Government), the Libyan National Guard, and/or local police agencies. The hope was that rolling the militia members into the new government was an appropriate way to thank them for their service. The government also hoped the militiamen would have careers and a purpose that they could continue to contribute their skills to a new Libya. At the end of the day, this policy failed, but they did not stop the funds that were going to terrorists.

I remember our Senior CIA Officer repeatedly explaining to the newly formed LIS that they needed to handle these terrorists immediately. It was handle them now, or a year from now, the country would have drones flying above it, turning into the tribal areas of Pakistan. They would try to walk back his talk, clearly thinking he was alarmist and told him that they had a grander plan for Libya and that he didn't understand that those in the revolution would prove to be patriotic to their government.

The ideals of how most Libyans wanted to establish their new democracy fell flat. And, sadly the Senior CIA Officer's projection came true within just months. I remember the first day when a drone flew overhead. While they were an important tool for finding terrorists like MBM, I remember looking up at them and wondering how it felt to be Libyan as it was flying there because terrorists were among them. Would they think that we thought they were the bad guys, that they had done something wrong in their fight against tyranny? I wanted access to that tool so bad then, but a decade later, I want more for the day Libya doesn't need drones, airstrikes, or counterterrorism operations.

We will focus on Benghazi, a snapshot, one moment in time, but we acknowledge to the Libyans there were numerous hotspots of conflict. Benghazi may not have been the most important to them even though, in our personal lives, it was the most important incident to us. We respect that others have struggled in Libya on a level that we cannot address within the bounds of our focus on our attackers. Also, as Americans, we cannot ignore that our involvement in Libya, while morally right and honorable to protect the city of Benghazi during the revolution, was also a failed U.S. policy that played a role in the terrorism that was to become—in Benghazi.

Terrorists forced their will on the Benghazi population through fear and force through tactics like assassinations. The early Benghazi assassination units were developed by Senior al-Qa'ida Operative Abu Anas al-Libi, with real name Nazih Abdul-Hamed Nabih al-Ruqai'i. While the revolution was supposed to belong to all Libyans, it no longer belonged to the citizens who had been the heart and soul of the revolution. It belonged to their new oppressors who melded their powerful militias into al-Qa'ida affiliates like Ansar al-Sharia.

And from the day it went up in October 2011, that damn black flag remained above the courthouse until the attack, and then as one could expect, a black flag was carried onto our Consulate that fateful night in September 2012.

Chapter 4

Local Political Dynamics

Politics in Benghazi, in particular the September 12th Prime Minister elections, played a key role as to why the militias noted as U.S. "allies", either refused to provide support, or were complicit in the six attacks over the course of September 11th and 12th, 2012

Local politics influenced how the militias, perceived as allies, either refused to provide support once the September 11th Benghazi attacks had started or decided just to attack us themselves. Does one wonder why our key allies in Benghazi were militias to begin with? This was easy to explain, as the Libyan government was in upheaval and kept changing. They could not transition enough power to an official government structure, allowing Libyan militias almost unlimited power and influence. Our Government leaned into trusting the militia model because that was the status quo at the time, but it was a big mistake.

Approximately 200,000 local Libyans joined a militia to support the rebellion. The militias were formed as a counterweight to Gaddafi's well-trained, well-armed, and what was believed to be a loyal army. Some militias were proximity based; some were based around an individual or were developed to honor a deceased national hero; some were formed to rectify a wrong among its participants; some were focused on secular political beliefs fighting for a democratic state; and some were hardcore terrorists fighting for the formation of an Islamic state under Sharia Law.

When fighting Gaddafi, everyone fought as one body; they were the Revolutionaries. After the fact, it's easy to put an asterisk on the "bad guys," but during the revolution, that did not exist—everyone was a hero to the Libyan people. It was only in the years after the revolution that the cleavages between everyday patriotic Libyan-loving individuals and terrorists who only cared about a Libya that suited their extremists' ideas would emerge. And honestly, even as these cleavages became apparent, there are still hardcore terrorists celebrated as heroes in Libya, and it's a difficult bond to break.

Did the Libyan Government not develop properly due to the militias, or did the militia's power grow to the failure of the Libyan Government to form effectively? Regardless of the causes and effects, within that first year of independence from Gaddafi before the Benghazi attacks, the Libyan Government had been incapable of bringing the country together primarily because of the legitimacy of the new Government referred to as the National Transitional Council (TNC) was not on solid ground. As background, the TNC was the temporary Government in Libya set up in Benghazi on February 27th, 2011, to be the people's Government. In reality, it was essentially the rebel government that had fought against Gaddafi.

Once Gaddafi was toppled, the initial plan had been for these militiamen and capable revolutionaries to be rolled into a new government. The problem was that many Libyan citizens could not hold power during Gaddafi's regime, protect themselves, nor be leaders among their fellow citizens. Those who had been severely disadvantaged under the Gaddafi regime were unwilling to cede control. Each time the TNC made a subsequent formal agreement offering authority to militias, it increased the militias' power and helped to destabilize the weak TNC. The Government had to appease various parties to garner support for their structure; these parties also included terrorists. As a result, they essentially contracted out much of their security and law enforcement-related duties to influential militia leaders under authority given to the Supreme Security Committee (SSC) and the Libya Shield Forces (LSF) instead of forming government bodies to perform those roles.

The leader of the board of the TNC during the revolution, which was equivalent to the title of Prime Minister, was Mahmud Jibril el-Warfally (known hereafter as Jibril). By the Summer of 2012, Libya was beginning to transition out of the TNC. They were attempting to form a new government rule that was still transitionary. It was coined the General National Congress (GNC), and it was given the authority to transition the Government into a form of legislation that would govern under a democratic constitution.

The popular vote to establish the GNC was held on July 7th, 2012. AAS-B's Leader Mohammad al-Zahawi forbid members from voting in the election. Separately, the U.S. and Ambassador Stevens were known to be staunch supporters of Jibril as a candidate for Prime Minister of the new GNC. Bringing Jibril back into the fold was reminiscent of when he and Ambassador Stevens worked closely together in Benghazi during the revolution.

In March 2011, Jibril had received intelligence that Gaddafi had intended to destroy Benghazi. The city was the rebels' stronghold and housed Gaddafi's inner circle who defected to the rebellion. With this intelligence in hand, Jibril petitioned French President Nicolas Sarkozy and Ambassador Stevens. Stevens then garnered the support of Secretary Clinton to push for NATO involvement.

Secretary Clinton then got President Barack Obama on board to protect Benghazi from a mass atrocity event and rid America of Gaddafi, who had a long history of undercutting western interests. Islamists in Benghazi were not at all happy with Jibril bringing the U.S. and NATO into the revolution. They believed they could defeat Gaddafi without western support and were concerned about how western influence would affect their aspirations for Libya to function as an Islamic state. In addition, for some segments of the revolutionaries, Jibril wasn't trusted because he was from the al-Warfalla tribe. The tribe had historically claimed a long allegiance to Gaddafi's rule.

Jibril was running against a bloc backed by Islamists and with some backing of the Muslim Brotherhood. The Libyan Muslim Brotherhood was a slightly muted version of the actual Muslim Brotherhood and, in 2012, was represented by the Justice and Construction Party (JCP). While there were some security incidents against polling places and protests during the July 7th Libyan elections, most residents in Benghazi were able to participate in the election.

The population of Benghazi was estimated to be about 500,000 total persons in 2012. During the election in Benghazi, Jibril's National Forces Alliance received just over 95,000 votes compared to slightly over 16,000 for the JCP. Jibril's coalition won 39 of 80 seats, and the JCP took only 17 seats. On August 8th, 2012, the GNC officially took control of the Libyan Government.

Then the next day, the GNC elected the National Front Party Leader Mohammed Yousef el-Magariaf as Libya's new President. Magarief, a "moderate Islamist," two words that make entirely no sense together, won 113 votes. On September 14th, 2012, just two days after the Benghazi attacks, in an interview with Al Jazeera, Magarief stated the following publicly: "I think this was al-Qa'ida," and he went on to say, "It's a dirty act of revenge that has nothing to do with religion."

Magarief also said the idea that the attack was a "spontaneous protest that just spun out of control is completely unfounded and preposterous." Magarief further noted on September 16th, 2012, that the attack on the U.S. Consulate in Benghazi was a planned terrorist attack. Another comment Magarief made in an interview regarding the U.S. was, "It may be better for them to stay away for a little while until we do what we have to do ourselves."

Unfortunately, the U.S.'s withdrawal from Benghazi allowed for leaders like Magarief to continue to appease Islamists, and the Libyan Government became more intertwined with the Islamists and then terrorists, not less. Even after the 2012 attacks, when the residents of Benghazi asked Magarief for a change, he held a press conference supporting militias.

An example of how hard it was for us to know friend from foe in Benghazi: GNC's Mohammad al-Magariaf's September 22nd, 2012 press conference announcing the formation of a new "Military" Committee in Benghazi; Included the following militia leaders associated with Benghazi attacks—Wissam bin Humaid, Salem Darby, Ahmed al-Majbari, Mohammad al-Gharabi and Fawzi bu Khatif; Also includes Mustafa al-Saqzli (the brother of al-Qa'ida attacker Khaled al-Saqzli)

The next phase of the elections was to choose a Prime Minister, the date for the elections was September 12th, 2012, the day after the attack on the U.S. Consulate. The newly elected 200 members of the GNC would choose the next Prime Minister. The success of Jibril's coalition in the general election concerned militias in Benghazi who had not expected his party to do so well in July. They were worried he would also lead the new Government of the GNC. Jibril was avidly anti-Islamists and wanted to keep them out of the government, especially out of the security, military, and police apparatuses that the militias had controlled since 2011.

On September 5th, 2012, the new GNC structure submitted nominees to replace Libyan Prime Minister Abdurrahim el-Keib, who served from November 24th, 2011, to November 14th, 2012. The primary candidates included Jibril, Awadh al-Barassi, and Mustafa Abu Shagur.

In terms of players in Benghazi, CIA Annex attack Mastermind, Wissam bin Humaid, and Rafallah al-Sahati Brigade and al-Qa'ida commander Mohammad al-Gharabi supported the JCP and were hoping to elect Awadh al-Barassi. Awadh was the main competitor to Jibril. According to Gharabi and Wissam, Barassi intended to appoint 17 February Leader Fawzi Bu Khatif as the Libyan Minister of Defense.

A new Prime Minister spelled uncertainty for the militias, particularly if Jibril won. So, on September 9th, 2012, Gharabi and Wissam made it clear to the Principal Officer from the U.S. Consulate (one day before Ambassador Stevens arrived in Benghazi) that they would not continue to guarantee security in Benghazi because the U.S. was supporting Jibril as a candidate for Prime Minister. Further, just one day prior, on September 8th, 2012, Fawzi Bu Khatif, whose militia was operating as the armed guards on the U.S. Consulate compound, told the U.S. Officials in Benghazi that the group would no longer support off compound moves. CIA ended up supporting these moves in place of 17 February.

And what was learned after the fact is not only did Libya Shield, 17 February, nor Rafallah al-Sahati Brigade fail to render aid when we were attacked, but all three groups provided terrorists to al-Qa'ida for the attacks. Then Wissam fired mortars at the CIA Annex to deal the last blow to the U.S. presence in Benghazi. These leaders were frank, they forecasted their intent not to support us, and we should have listened.

On September 12th, Jibril was defeated in the Prime Minister contest by Mustafa Abu Shagur by just two votes. The GNC was dissolved on April 1st, 2016. Years later, Jibril suffered from cardiac arrest, later tested positive for COVID-19, and died on April 5th, 2020. And no, Fawzi bu Khatif never became the Minister of Defense for Libya.

Chapter 5

Enemies: The Masterminds

"The Masterminds" encompasses the senior most plotters of the U.S. Consulate in Benghazi and CIA Annex attacks in September 2012.

Included herein are: (1) Dr. Ayman al-Zawahiri, al-Qa'ida's former Leader who directed the attack; (2) Abdel Malek Droukdal, al-Qa'ida in the Islamic Maghreb's (AQIM) Leader in 2012 who tasked the kidnapping plot; (3) Mokhtar Belmokhtar, AQIM Battalion Commander who was the Mastermind of the U.S Consulate attack; (4) Wissam Bin Humaid, Libya Shield One Leader who was the Mastermind of the CIA Annex attack, and (5) Omar al-Shalaali, AQIM's Leader for Libya who was the head al-Qa'ida Commander on the ground directing the attacks on the U.S. Consulate.

(1) Dr. Ayman al-Zawahiri, full name Ayman Mohammed Rabie al-Zawahiri from Egypt. Until the CIA killed him in a drone strike in Kabul, Afghanistan, on July 31st, 2022, Zawahiri was the Leader of al-Qa'ida. He replaced Osama bin Laden, who was killed in a raid led by the CIA and U.S. Navy SEALs on May 2nd, 2011.

Zawahiri began his journey into radicalism during his college years. He became involved with the Islamist political organization the Muslim Brotherhood and soon began advocating for the use of violence to achieve their goals. He later studied medicine but left his medical career behind to join the EIJ. He quickly rose through the ranks of the organization, and he played a key role in the assassination of Egyptian President Anwar Sadat in 1981. Zawahiri had been involved in terrorist activity for most of his adult life and had been arrested and imprisoned several times by Egyptian authorities.

EIJ is pertinent to the story of the U.S. Consulate attacks in Benghazi. Zawahiri and his brother Mohammad al-Zawahiri, full name

Mohammad Rabie al-Zawahiri, reached out to their prior networks within EIJ to support the attacks. Two prominent former EIJ members asked to support the attacks included Murjan Salim and Mohammad Jamal, whom Mohammad al-Zawahiri served with in Egyptian prison until the Arab Spring. At the time of the Benghazi attacks in 2012, Jamal had founded al-Qa'ida in Egypt (AQE).

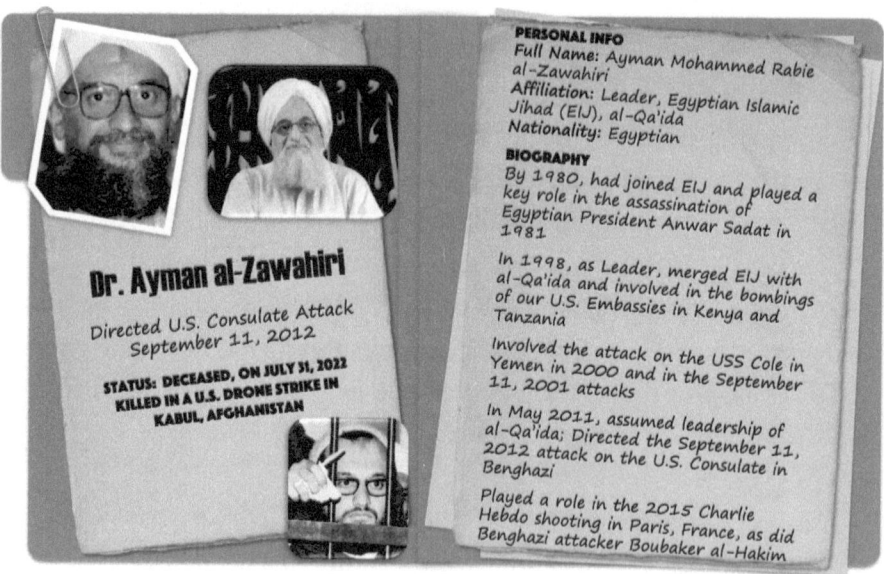

Dr. Ayman al-Zawahiri
Directed U.S. Consulate Attack
September 11, 2012

STATUS: DECEASED, ON JULY 31, 2022
KILLED IN A U.S. DRONE STRIKE IN
KABUL, AFGHANISTAN

PERSONAL INFO
Full Name: Ayman Mohammed Rabie al-Zawahiri
Affiliation: Leader, Egyptian Islamic Jihad (EIJ), al-Qa'ida
Nationality: Egyptian

BIOGRAPHY
By 1980, had joined EIJ and played a key role in the assassination of Egyptian President Anwar Sadat in 1981

In 1998, as Leader, merged EIJ with al-Qa'ida and involved in the bombings of our U.S. Embassies in Kenya and Tanzania

Involved the attack on the USS Cole in Yemen in 2000 and in the September 11, 2001 attacks

In May 2011, assumed leadership of al-Qa'ida; Directed the September 11, 2012 attack on the U.S. Consulate in Benghazi

Played a role in the 2015 Charlie Hebdo shooting in Paris, France, as did Benghazi attacker Boubaker al-Hakim

This network of individuals also supported the unrest at the U.S. Embassy in Cairo, Egypt, on September 11th, 2012, starting approximately five hours before the Benghazi attacks. With financial support from the Leader of AQAP, Nasir al-Wuhayshi, Jamal and Murjan sent fighters to participate in the Consulate attacks. The attackers they sent were from an AQE cell coined the Nasr City Cell, and Egyptian authorities carried out several raids on the cell in the aftermath of the Benghazi attacks. The cell was also planning follow-on Embassy and Consulate attacks, including in Egypt, with former al-Qa'ida in Iran Leader, and Khorasan Group Leader Muhsin al-Fadhli.

Apart from AQE, a large number of Benghazi attackers had been prior members of EIJ, including Ali al-Karshini, Abu Hamza al-Tabawi, Ahmed Abu Khatallah, Jaafar Azzouz, Khaled al-Ammari, Atef al-

Karami, Hamad al-Fakhri, Ahmed al-Munfi, Marei al-Manfi, Abdullah Bouzkia, Talal bin Hariz, Salah al-Mushaiti, Shoaib al-Mushaiti, Khaled al-Faydi, and Youssef al-Darsi. A third set of Egyptian Benghazi attackers included Asmi Ahmed, Ahmed Hazaa, Karim Moawad al-Rahmani, and Mohammad Jaber Abd al-Maqsoud. These terrorists participated in the attack on the U.S. Consulate in September 2012 with Malik al-Khazmi and in the early 2000s were part of an Ansar Allah cell. In 2006, all five were arrested for participating in this terrorist cell and were sent by the Gaddafi regime to Abu Salim prison in Tripoli.

Zawahiri first arrived in Pakistan in the 1980s, during the Soviet-Afghan War. The EIJ members mentioned earlier, Murjan Salim and Mohammad Jamal followed Zawahiri to the Afghanistan and Pakistan border region. Zawahiri initially provided medical assistance to the mujahideen fighting the Soviet Army in Pakistan. He then went over to Afghanistan, where he met Osama Bin Laden. At the time, bin Laden was running the Khaldan Camp for mujahedeen fighters who were battle-hardened from the war against the Soviets. Zawahiri, who had also fought in that war, was impressed with Bin Laden and his fighters. They shared a common goal to rid the Middle East of American influence, and soon became fast friends.

Other senior leaders involved in the Benghazi attacks, which Zawahiri met in Afghanistan, included Abdel Malek Droukdal, the Leader of AQIM; Mokhtar Belmokhtar (MBM), a Senior Battalion Commander in AQIM and the Mastermind of the Benghazi attacks; Mohammed al-Gharabi, the Senior al-Qa'ida Commander in Benghazi in 2012; and Seifallah Ben Hassine, the Leader of Ansar al-Sharia-Tunisia (AAS-T). When Zawahiri moved up into the senior ranks of al-Qa'ida, he had a close and collaborative operational relationship with Droukdal, where Zawahiri could task AQIM, and Droukdal could weigh in on senior-level group decisions for core al-Qa'ida. Zawahiri also met the following Benghazi attackers in the Pakistan and Afghanistan border region: Marei Zoghbi, Faraj al-Chalabi, and Hamza al-Darnawi.

Zawahiri would join Bin Laden's side in his fight against the West. Together, they led al-Qa'ida after the EIJ merged with the group in 1998,

Enemies: The Masterminds

making Zawahiri second-in-command of the terrorist organization. From the base in the Pakistan and Afghanistan border region, Zawahiri would help plan and carry out some of the most devastating attacks in history. In addition to directing the U.S. Consulate in Benghazi attacks on September 11th, 2012, Zawahiri was believed to have played a role in the 2015 Charlie Hebdo shooting in Paris, France, an attack that Benghazi attacker Boubaker al-Hakim was also involved in; the 2001 September 11th attacks on the United States; the 2000 attack on the USS Cole in Yemen; and the bombings of our U.S. Embassies in Kenya and Tanzania in 1998.

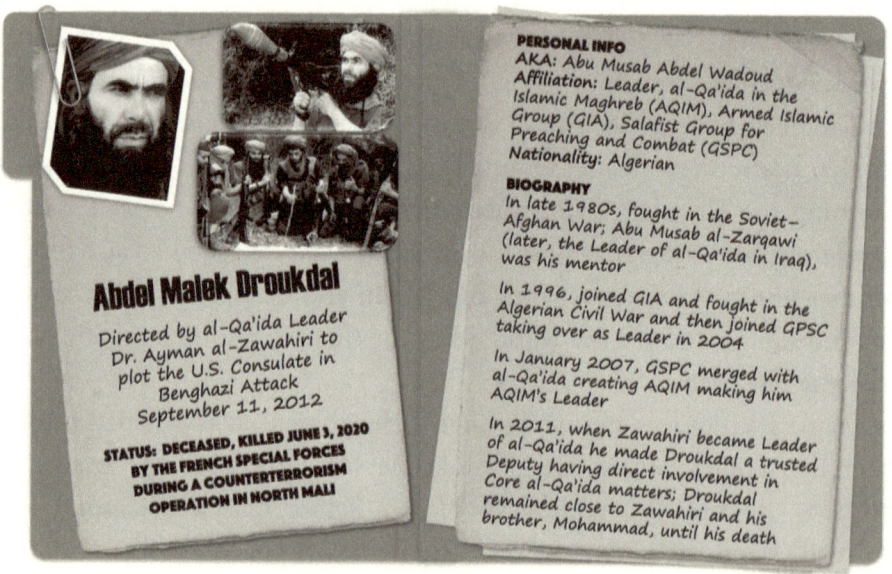

(2) Abdel Malek Droukdal with alias Abu Musab Abdel Wadoud from Algeria. At the time of the Benghazi attacks, he was the leader of AQIM. Nearing the end of the Libyan revolution in 2011, Droukdal was keen to take advantage of the security vacuum in Libya to establish a safe haven for AQIM in the country. Unlike core al-Qa'ida, which focused on major population centers in Libya like Tripoli and Benghazi for safe havens, Droukdal focused on more remote desert regions. As Droukdal had success during this post-Gaddafi timeframe, he was directed by al-Qa'ida Leader Dr. Ayman al-Zawahiri to attack the U.S. Consulate in Benghazi.

In addition to leading AQIM, Droukdal had an advisory position with Zawahiri; essentially, he could provide input and was involved in Zawahiri's decision-making when it related to all of al-Qa'ida's global affiliates.

In the 1980s, Droukdal became a close associate of Zawahiri when he joined al-Qa'ida after traveling to Afghanistan to fight in the Soviet-Afghan war. In Afghanistan, his mentor was Abu Musab al-Zarqawi, the founder of both al-Qa'ida in Iraq and ISIS. During the 1990s, Droukdal returned to Algeria and fought in the Algerian Civil War, first joining the Armed Islamic Group (GIA) and then joining its offshoot Salafist Group for Preaching and Combat (GSPC) as a Senior Leader. In June 2004, Droukdal became GSPC's Leader after its Leader Nabil Sahraoui alias Mustapha Abou Ibrahim was killed in an Algerian Army counterterrorism operation.

In 2007, GPSC merged with AQIM, with Droukdal and MBM vying to be chosen as leaders. Both had served in Afghanistan with Osama bin Laden and Zawahiri, but Droukdal won in the end. This appointment caused a constant conflict between the two senior leaders. This split was evident during the September 11th, 2012, attacks on the U.S. Consulate: 1. While MBM was the attack Mastermind, other brigades within AQIM were deployed to participate in the attacks to include AQIM's Mali Group; 2. Droukdal maintained at least two separate reporting structures during the attacks; and 3. The AQIM Leader for Libya, Omar al-Shalaali, who was not a member of Belmokhtar's al-Mulathameen Battalion, led the attack on the ground.

As an aside, MBM led much of the attack planning efforts with AAS-B. Droukdal received feedback in real-time from Omar al-Shalaali that AAS-B was underperforming during the U.S. Consulate attacks. As such, the AAS-B attackers that had been slotted to participate in the In Amenas, Algeria attack in January 2013, were removed from the plot. It was unclear if MBM agreed with this position, but he likely did not, as he continued a close relationship until his death with AAS-B. It also showed that MBM continued to consult with Droukdal on key operational decisions even as there were rumors throughout most of 2012 of their permanent split from one another.

On June 3rd, 2020, French Special Forces killed Droukdal during the

Battle of Talahandakin, Mali. Abu Ubaidah Youssef al-Annabi succeeded him. Of interest, it was both Droukdal and Annabi who in October 2015, sent out an international plea for terrorists to respond to Benghazi to help save al-Qa'ida's primary base in Ganfouda during the Battle of Benghazi. In their request, they also evoked the memory of Abu Yahya al-Libi, a huge rallying cry during the September 11th, 2012, Benghazi attacks. Their appeal temporarily worked, as enough terrorists filtered into Benghazi for the battle to continue for almost 14 more months.

The Battle of Benghazi finally ended three weeks after the early December 2016 death of the al-Qa'ida golden child and CIA Annex attack Mastermind, Wissam bin Humaid. Al-Qa'ida feared the death of Wissam would signify the end of the war. To keep terrorists focused on fighting the Libyan National Army (LNA), al-Qa'ida worked hard to cover up Wissam's death; however, the ploy did not buy al-Qa'ida much time. By late January 2017, the LNA defeated al-Qa'ida's base in Benghazi and had killed many key al-Qa'ida-affiliated senior terrorist leaders.

To walk through a little-known fact, besides killing so many al-Qa'ida terrorists from 2014 to 2017, when the LNA kicked off its Operation Dignity in the Battle of Benghazi in 2014—it also defeated Droukdal's near-term plans at the time. Droukdal fought hard to help al-Qa'ida keep its base, which included directing two Benghazi Consulate attackers, Anis al-Houthi and Mahmoud al-Wahishi, to command a large number of AQIM-affiliated terrorists from Algeria, Tunisia, and Yemen in the Battle of Benghazi.

By 2014, Droukdal had been coming to terms with the fact that he could no longer stay in the Kabylie Mountains of Northern Algeria as his safe haven was slipping away as the Algerian Army continued to make gains against the group. He debated relocating to Libya and Tunisia; however, the LNA ensured that Libya was no longer a viable option for the leader. Preparing for a future without Benghazi, Droukdal was wise not to move all his resources in Libya to the front and stayed focused on other locations that could help AQIM maintain a safe haven. For MBM that was Ajdabiya, Libya. For AQIM, writ large, a key place, still a safe haven for the group in 2022, was the Tendi Mountains in southern Libya.

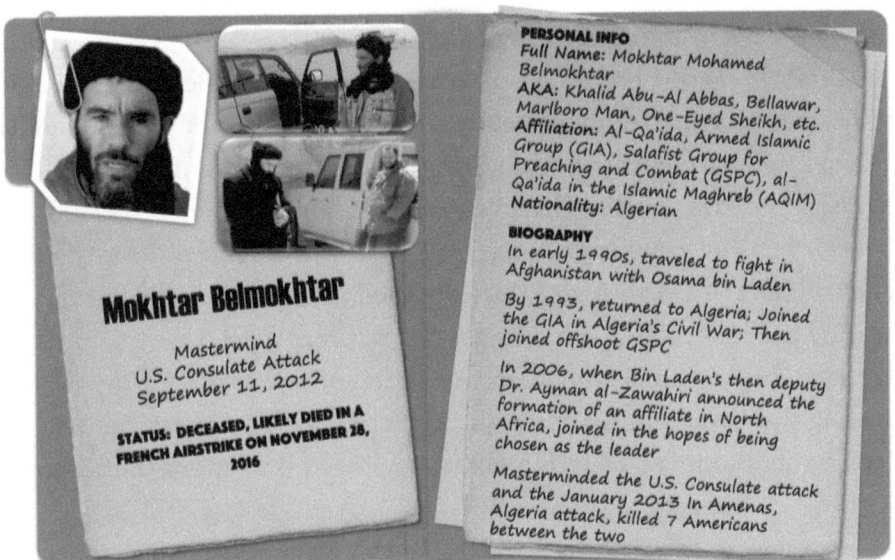

(3) **Mokhtar Belmokhtar** (MBM), full name Mokhtar Mohamed Belmokhtar from Ghardaia, Algeria. At the time of the September 2012 Benghazi attacks, MBM was a member of AQIM and ran its al-Mulathameen Battalion, one of several Battalions in AQIM's Saharan Brigades. MBM had many aliases, with the primary ones being Abu-Al Abbas, Bellawar, Marlboro Man, Le Borgne, and One-Eyed Sheikh.

Even as MBM was known to butt heads with AQIM's leader Abdel Malek Droukdal and within AQIM writ-large, which led him to split from the group eventually, he always remained al-Qa'ida. No matter the name of the group or Battalion that MBM was associated with at any given time nor what city or country he was located in, he always viewed Osama bin Laden and Dr. Ayman al-Zawahiri as his Leaders.

In 1991, when MBM was just nineteen years old, he traveled to Afghanistan to train in al-Qa'ida camps initially established to support the Soviet-Afghan War, which lasted from 1979 to 1989. He fought alongside Osama bin Laden, and his time with Bin Laden meant so much to him that he named a son, Osama, after the al-Qa'ida leader. MBM famously lost an eye, similar to Afghan Taliban Leader Mullah Mohammad Omar Mujahid Akhund (known hereafter as Mullah Omar), while fighting in Afghanistan. While still in Afghanistan, back home in Algeria, the Algerian Civil War kicked off in December 1991.

By 1993, MBM had returned to Algeria to fight in the war and joined the Armed Islamic Group of Algeria. He quickly rose to be a Commander due to his jihadi credentials from being al-Qa'ida in Afghanistan. In 1998, MBM joined the Salafist Group for Preaching and Combat (GSPC). In 2006, when Bin Laden's then deputy Zawahiri announced the formation of an affiliate in North Africa, MBM joined in the hopes of being chosen as the Leader of AQIM. He was passed over for Abdel Malek Droukdal. He later broke the al-Mulathameen Battalion off from AQIM. While leading the battalion, his forte was kidnappings for ransom operations, but he was also involved in human smuggling and weapons trafficking as it's good to diversify.

Now on to Libya. After the 2011 Libya revolution, MBM was keen to take advantage of the security vacuum. He first went to Libya and was stationed in the Tendi Mountains near Ubari. This location offered numerous opportunities. First, AQIM had relationships with local tribes, which helped ensure their security. In addition, it was a primary thoroughfare to move smuggled goods across Libya. With Ubari as a hospitable base, MBM focused on outreach toward the coast. He was known to be able to establish trusting relationships quickly and formed alliances with organizations in Ajdabiya, Al-Nawfaliyah, Benghazi, Darnah, and Sirte. Most of his relationships exploited extremist and smuggler networks.

To narrow the scope, as MBM had a long jihadist history, MBM was the Mastermind of three high-profile attacks in North Africa starting within a year of September 2012.

1. On September 11th, 2012, MBM was the Mastermind of the al-Qa'ida plot to kidnap Ambassador Stevens at the U.S. Consulate in Benghazi.
2. From January 16th to 19th, 2013, MBM was the Mastermind of the In Amenas Oil Facility Attack and Hostage Crisis. This terrorist incident killed 37 hostages from 9 countries, including 3 Americans. Over 35% of MBM's In Amenas attackers were the same attackers from the U.S. Consulate in Benghazi.

3. On May 23rd, 2013, MBM was the Mastermind of two simultaneous attacks. The first one was a vehicle-borne improvised explosive device (VBIED) attack on Areva, a French-owned uranium mine in Arlit, Niger. This terrorist incident killed one employee and injured 13 others. The second one was a VBIED attack on a Nigerien Military base in Agadez, Niger. This terrorist incident killed 26 and injured 30.

In August 2013, MBM merged his al-Mulathameen Battalion with al-Mourabitoun, another AQIM offshoot that rolled back into AQIM in 2015 when it reported MBM was its Leader at that time. What is important to note is that MBM was a co-founder of the al-Mourabitoun group in Darnah, which in true MBM fashion, reported to Zawahiri and not through AQIM's al-Mourabitoun. MBM personally fought in the al-Qa'ida battles of Darnah against ISIS to ensure al-Qa'ida controlled Darnah. In August 2015, ISIS listed MBM as one of its most wanted terrorists in Darnah.

MBM founded al-Mourabitoun in Darnah with Egyptian terrorist Hesham Ashmawy, full name Hesham Ali Ashmawy Mos'ad Ibrahim with alias Abu Omar al-Mujhajir. The group was founded in alliance with several other high-profile regional terrorists, including Omar Rifai Sorour alias Abu Abdullah al-Masri who was the Mufti of the Darnah Mujahideen Shura Council (DMSC); Mohammad Fathi Karim with alias Abu Malik; and Bahaa Ali Ali Abu al-Maati with alias Abu Abd al-Rahman. On June 9th, 2018, Omar Rifai Sorour was killed with Benghazi Consulate attacker Abu Zaid al-Shalawi in an LNA airstrike. Omar's father, Rifai Sorour, was a prominent Egyptian jihadi Salafist who was a close associate of al-Qa'ida Leader Dr. Ayman al-Zawahiri.

On October 8th, 2018, during a counterterrorism operation in Darnah, LNA captured Ashmawy, Bahaa Ali, and Consulate attacker Marei Zoghbi (also affiliated with al-Qa'ida's Milan Cell in the past). On May 28th, 2019, the LNA handed Ashmawy and Bahaa Ali over to Egyptian authorities after negotiations for the two high-value terrorists. Ashmawy was charged for his role in 54 separate terrorist operations.

Egypt executed both, with Ashmawy's execution occurring on March 4th, 2020. As of 2022, Consulate attacker Zoghbi remained in LNA custody in Libya.

MBM spent extensive periods in Ajdabiya, whereby in 2012, he had a trusted network that provided a safe haven for him in the city. His primary terrorist associates in Ajdabiya were Marei Jbeil, Abdel Qader Fneish al-Zawi, Al-Kilani Bounawara, and Al-Saadi al-Nawfali with real name Al-Saadi Abdullah Ibrahim Bukhazem. In a circa 2015 video, MBM was seen with all these Ajdabiya associates, many of whom were members of the al-Qa'ida-affiliated Ajdabiya Revolutionaries' Shura Council.

On June 15th, 2015, the U.S. conducted an airstrike on Nawfali's farm on the outskirts of Ajdabiya in an attempt to kill MBM. Seven terrorists were killed affiliated with AAS-B, including Attiya Bouhadeeda al-Mughrabi, Naji Bouhadeeda al-Mughrabi, Said Boukhazim al-Mughrabi, Ahmed Khattab al-Fakhri, Ali Badi, Hamza al-Mushaiti and Faraj Doma al-Zawi. AAS-B issued a statement mourning these terrorists but denied the killing of MBM. At the time, the Libyan Government also confirmed his death. The incident indicated that MBM maintained close ties to AAS-B since the 2012 Consulate attacks.

There was a joke in intelligence circles that MBM was killed every year in an airstrike. This air-raid was no exception. After the failed attempt, MBM assisted in planning the November 20th, 2015 Bamako hotel attack in Mali killing 20. Another American and former Peace Corps volunteer Anita Ashok Datar from Takoma Park, Maryland was killed, as well. On November 28th, 2016, the French conducted an airstrike based on an information-sharing relationship with the U.S. that killed MBM. MBM was killed two days after the successful LNA strike on the CIA Annex Mastermind, Wissam bin Humaid. In 2017, the leadership ranks of the greater al-Mourabitoun group in Mali formally removed MBM from its roster The group's action acknowledged that the senior leader's nine lives were finally over.

(4) Wissam bin Humaid, full name Wissam Faraj Mahmoud bin Humaid with alias Abu al-Khattab al-Libi from Benghazi. Wissam catapulted to fame during the Libyan revolution in 2011 when he was the Leader of the Free Libya Martyrs Brigade. At the time of the attacks, Wissam was the Leader of Libya Shield One, and he was the Commander of several camps run by the organization in Benghazi.

Wissam, a member of al-Qa'ida, then pledged allegiance to ISIS and fought with the group and AAS-B against the LNA. Wissam died a member of al-Qa'ida while fighting to maintain the group's primary Libyan base in Ganfouda, Benghazi, Libya. The ISIS and al-Qa'ida divide did not occur immediately in Libya, so many terrorists pledged allegiance and went to fight with ISIS when it created its local branches, with Wissam pledging allegiance to the group in 2015. However, terrorists then had to choose a side. When these two groups turned on each other, some chose al-Qa'ida, and some chose ISIS—in Wissam's case, he stayed loyal to al-Qa'ida and the group to him—but both groups wanted to claim him even in death.

Wissam was like a God in Libya, all presumed to be due to his role in the revolution. Still, in secret (inside terrorist groups), it was due to him being the Mastermind of the CIA Annex attack on September 12th, 2012. Wissam took advantage of the al-Qa'ida attack on the U.S.

Enemies: The Masterminds

Consulate to strike the Annex. The AQSL plot only involved the U.S. Consulate with Ambassador Stevens as its target. Wissam saw the chaos of the evening and realized it was his opportunity to rid the city of the Americans for good.

On September 11th, 2012, while canvassing the scene outside the U.S. Consulate immediately in the initial attacks' aftermath, Wissam received information likely from Boka al-Oraibi, the Leader of Libya Shield Two, that the Americans were sending a rescue force from Tripoli, coined "Team Tripoli." Wissam was enraged that even more Americans would be coming to the city, and he verbally vowed to kill all of us that evening. Wissam likely collaborated with Boka, members of Libya Shield, and then used his two brothers, Mohammad and Qais bin Humaid, to orchestrate the timing of the mortar strike on the morning of September 12th.

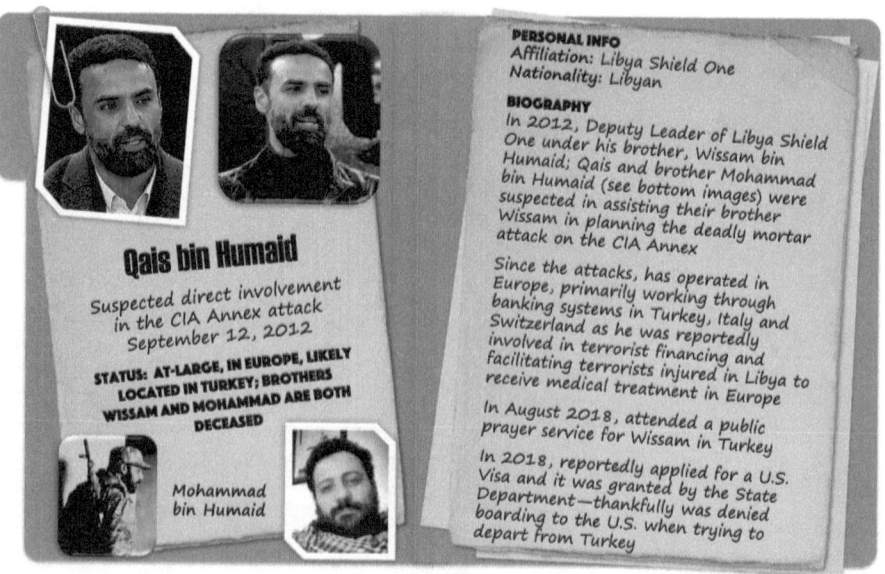

Mohammad with an alias that translates into English as "the Gulf Sniper," was a fighter who the LNA killed in the Battle of Benghazi. In contrast, Qais bin Humaid was the one who got the finances together to fund Wissam's terrorist activities. As of 2022, Qais was still at-large, likely located in Turkey, and reportedly continued his involvement with terrorist financing for the last decade using networks in Europe, primarily in Italy, Switzerland, and Turkey. When we were in Libya in 2012, Qais, along with his brother, Wissam, would reportedly meet with the interim Libyan Prime Minister Abdurrahim el-Keib to secure funds from the Government of Libya to support terrorists in Benghazi.

Back to the evening of September 11th, 2012, one of the moving parts that Wissam needed to control as he planned his deadly attack on the CIA Annex was the need to delay "Team Tripoli" from leaving the Benghazi International Airport upon arrival. Wissam needed to ensure he could simultaneously get all the Americans collocated inside the Annex. To do this, Wissam used Boka and his Libya Shield Two militia to help align the assault on the CIA Annex. While Boka was witting, it is assessed the two used one of his Commanders, Fathi al-Obeidi, to support this initiative unwittingly. It does not appear that the ground unit Fathi commanded was witting either. Not only was Wissam successful in secretly carrying out the mortar strike on the CIA Annex, State Department still reports that the Americans in Benghazi were rescued from the Annex by Libya Shield. And it was Wissam who provided security for the FBI visit in October 2012, when they finally arrived in Benghazi to investigate the attacks.

Over the years, Wissam had a hand in countless terrorist incidents. The most prominent ones to highlight are:

1. On July 28th, 2011, he participated in the assassination of Major General Abdul Fatah Younis al-Obeidi with his brother Mohammad, and with terrorist Mustafa al-Raba and several Consulate Benghazi attackers, including Ahmed Abu Khatallah, Jamaica, Shoaib al-Mushaiti, and Zubayr al-Bakoush.
2. On June 8th, 2013, he led the Black Saturday Massacre in front of the Libya Shield headquarters, where terrorists opened fire on protestors, killing over 40. Other U.S. Consulate attackers were involved in the massacre to include Mohamed Ben Dardaf and Emad al-Shuqabi.
3. On April 15th, 2014, Wissam helped orchestrate the transfer of kidnapped Jordanian Ambassador, Fawaz al-Aitan, from Tripoli to

Benghazi. One of the locations where he detained the Ambassador in Benghazi was at the family home of Consulate attacker Ali al-Karshini. Consulate attacker Ayman Bouamoud carried out the kidnapping with the brothers of detained Libyan terrorist Mohamed el-Dresi variant al-Darsi with alias Mohammad al-Nass. In the end, the kidnapping was to gain the release of Dresi, who was exchanged for the Ambassador. Dresi was a Libyan AQI leader detained in 2007 and serving a life sentence in Jordan for an attempted AQI suicide bombing at the Alia International Airport in Amman, Jordan.

After Dresi's release, he, Wissam, and several terrorist associates, including Salem Jaber, Yahya al-Maqsabi, Ahmed Abu Khatallah, and many others, formed the al-Qa'ida-affiliated Benghazi Revolutionaries Shura Council (BRSC). At the time of his death, Wissam was married to the daughter of Jaber, who resided part of the time in Qatar and part of the time in Turkey as he has Qatari citizenship. Jaber was a Council of Senior Scholars member, an organization that was established in Switzerland and then moved to Qatar. The Council served as a front organization to advocate global support for extremist organizations.

Former al-Qa'ida Leader Osama bin Laden directed the formation of BRSC, years before it was actually established. It is commonly believed that BRSC was formed to fight the LNA, which was its main charge; however, that was not the group's origins. Just before his death in May 2011, Bin Laden realized that al-Qa'ida and al-Qa'ida-affiliated groups needed to coalesce in Libya because if the groups remained fractured and carrying different flags, al-Qa'ida would fail to establish a solid base there. Further, the groups would be more potent as a whole against the Gaddafi regime than in individual groupings and militias. Al-Qa'ida was at the forefront of global terrorist groups at the time, who guided foreign fighters to travel to Libya to support the revolution. As such, in 2011, he provided direction to the al-Qa'ida leadership in Benghazi, requesting the formation of shura councils to consolidate al-Qa'ida's power.

When al-Qa'ida members finally got around to forming the councils in 2014, Bin Laden's order led to the formation of BRSC and eventually the Darnah Revolutionaries Shura Council (DRSC) and the Ajdabiya Revolutionaries Shura Council (ARSC). Wissam was also a co-founder of ARSC. All three groups reported their activities to Bin Laden's replacement by this time, Zawahiri and received guidance from AQSL. AQIM's Droukdal also supported the BRSC model and aligned with the group to promote greater AQSL goals across North Africa. An added concern for AQSL was not just the counterterrorism operations LNA was ramping up that proved to be disastrous for the group, but whether the influence of ISIS would slow al-Qa'ida's growing base in Libya.

As an aside, while we try to stay focused on al-Qa'ida and its involvement in the attacks, the Palestinian terrorist group Hamas also had a relationship with some of the U.S. Consulate attackers and AAS-B members in Benghazi, including Wissam, who was in the "Hamas Cell." The Hamas Cell focused on smuggling weapons from Libya to Egypt and the Gaza Strip. The key terrorists affiliated with our story who were also members of the Hamas Cell included Wissam, Ahmed Abu Khatallah, Houjeen al-Mahdawi, Tariq Darman, Abu Dhar al-Saghir, Al-Tunisi al-Zawari, but there were many others. Khatallah reportedly had been detained briefly in Gaza in either 2011 or 2012. In 2016, the Israelis targeted Al-Tunisi al-Zawari for his role in this network. A key liaison from Hamas to Libya at the time was Marwan al-Ashqar. The relationship was so established that Marwan had a legitimate identification card issued by the Libyan National Guard, which suggests that some of the illegal weapon's shipments out of Libya to Hamas used the top cover of the Libyan National Guard for transport.

The FBI never added Wissam to the most wanted terrorist list, nor was he added to State Department's Rewards for Justice program. General Khalifa Haftar and his army, coined the LNA, have proven to be the lions of justice in our story as they made immeasurable sacrifices to hunt down the terrorists at the U.S. Consulate in September 2012. The initial war between the terrorists and the LNA was the Battle of Benghazi, which kicked off in May 2014. Wissam led BRSC to help defend the al-Qa'ida base in Ganfouda and the terrorist strongholds throughout various neighborhoods in Benghazi. In the battles to liberate Ganfouda in 2016, Wissam died after being seriously injured.

Specifically, on November 26th, 2016, the LNA with its Emirati allies targeted Wissam with a mortar round delivered via a drone while he rode in an armored vehicle. According to its diplomatic plates, this vehicle belonged to the Government of Sweden. Another vehicle occupant who died due to the mortar strike was AAS-B founder Jaafar Mohammad Bayou. Jaafar was the son of terrorist Mohammad Omar Hassan Bayou, who historically was one of the senior leaders of Ansar Allah. The strike did not immediately kill Wissam, and he was transported via bulldozer to the al-Hout Market in the al-Sabri neighborhood in Benghazi. On December 5th, 2016, Wissam succumbed to his injuries.

AQSL, including AQIM and BRSC, had all worked to conceal Wissam's death as they knew it would signify al-Qa'ida's defeat in Benghazi, with even ISIS publicly calling BRSC out for not reporting the death. In April 2018, Wissam's parents and brother, Qais bin Humaid, held a press conference to announce his death officially. After the formal announcement, key groups and individuals contacted the family to offer condolences. A few contacts to note were Huthaifa Azzam, the son of Abdullah Azzam, "the father of Jihad" who is best known as being the terrorist mentor to Osama bin Laden; the leadership of Hamas; and Khalid al-Sharif, the former Commander of the Military Wing of the LIFG, and the former head of the Libyan National Guard in 2012.

Separately, there were numerous obituaries produced by Wissam's allies and admirers, some of which noted that he had fought in Afghanistan and in Iraq with AQI. We could not find any evidence of him located in either country, so we do not believe he fought in either locale. Wissam did have a close personal relationship, though, with three members from al-Qa'ida who fought in Afghanistan that are worth noting, which included Abdel Moneim al-Madhouni (with Uwra al-Libi being the alias Bin Laden used for Madhouni), Nasser al-Qatous, and Benghazi Consulate attacker Marei Zoghbi. Wissam and Boka al-Oraibi remained heroes in Libya, and images of them are seen on vehicles and buildings both in

Benghazi and in the Libyan Government controlled areas of Tripoli.

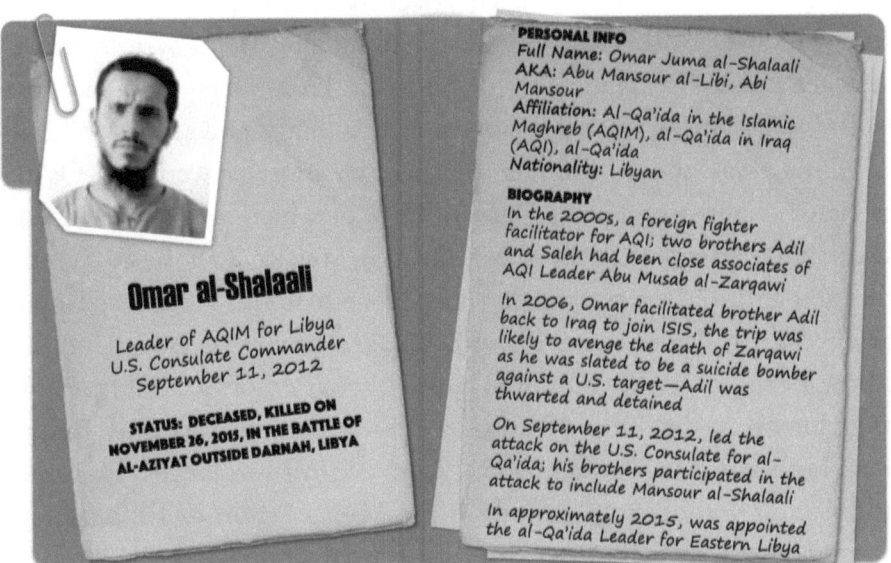

(5) Omar al-Shaalali, full name Omar Juma Mohammad al-Shaalali and aliases Abu Mansour al-Libi (from his time in Algeria) and Abi Mansour, from the eastern Shiha district in Darnah. Omar was reported to have used the alias "Abu Musab al-Mansour al-Libi" on the night of the Benghazi attacks. Omar is from a family of al-Qa'ida terrorists, with seven brothers and three cousins in total, five of which died in Iraq as members of AQI. At the time of the U.S. Consulate attacks on September 11th, 2012, Omar was AQIM's Leader for Libya. He also was the al-Qa'ida Commander who led the attack on the Consulate. When Omar died in 2015, he had risen to the rank of the al-Qa'ida Leader for Eastern Libya.

Omar joined AQIM in the early 2000s. In 2005, he recruited young men from Darnah, sent them to the mountainside in Algeria, and then facilitated them to Iraq via Syria to fight U.S. forces in Iraq. During this time, he became part of a terrorist cell in Libya's capital of Tripoli called the Rasheed Hotel Cell. Omar was arrested and jailed in Abu Salim prison for his involvement in the cell. He was freed from prison as the Libyan revolution kicked off.

During the revolution, on February 21st, 2011, Omar led an attack and captured the al-Abraq Air Base at the al-Baydah International Airport. He led several al-Qa'ida-affiliated attackers on the Air Base, including Bounqat al-Qabaili, Alaa al-Hasadi, and fellow Benghazi Consulate attacker and al-Qa'ida member Amin Kelfa. From 2011 to 2014, Omar led assassinations in Darnah against police, military, and security officials.

Omar was the co-founder of the two most powerful al-Qa'ida-affiliated terrorist groups in Darnah, both of which sent terrorists from Darnah to participate in the U.S. Consulate attacks. First, he co-founded the Abu Salim Martyrs Brigade (ASMB) with Abdul Hakim al-Hasadi, Salem Darby, Abdul Azim bin Ali, Bujaana with alias Aburashid, and fellow Benghazi attacker Abdel-Qadir Azzouz. As of 2022, Hasadi was a diplomat at the Libyan Embassy in Malaysia; Darby and Bujaana were both killed in Darnah; Abdul Azim bin Ali was located in Turkey; and Azzouz was likely in custody at the Mitiga Prison in Tripoli, Libya. Second, he co-founded Ansar al-Sharia-Darnah (AAS-D) with Sufyan bin Qumo. On June 9th, 2018, Qumo was reported to have been captured or killed during the LNA airstrike on Egyptian al-Qa'ida leader, Omar Sorour; however, contradictory reporting is available that suggests he may be at-large as of 2022.

When Omar led the attack on the Consulate on September 11th, his brother Mansour with aliases Al-Muhajir al-Darnawi and Abu Haroun, joined him. Mansour, also a member of AQIM, fought in Algeria from 2005 until 2011. At the start of the Libyan revolution, Mansour returned to Libya and fought Gaddafi forces. In 2012, he became a Senior Leader in Ansar al-Sharia-Benghazi's (AAS-B) umbrella organization and also was a senior operative in AQIM's Mali Group. On October 25th, 2015, AAS-B reported the death of Mansour.

Their oldest brother Mustafa, a former Abu Salim prisoner like Omar, was also suspected of being involved in the Consulate attacks. He fled to join AQI in 2005 and was detained and imprisoned in Iraq. In 2009, the Gaddafi Foundation secured his release. After the Libyan revolution, Mustafa returned to al-Qa'ida-affiliated terrorist activities and, as of 2022, had gone off the grid. A fourth brother, Saleh, was a member of AQI and killed in 2005 by the U.S. Military near Fallujah, Iraq. The fifth and sixth brothers, Fathi and Mohammad, also suspected of being involved in the Consulate attacks, were AAS-D members in September 2012. As of 2022, Fathi was at-large. Mohammad was an arms smuggler across the southern border of Libya. As of 2022, he maintained a low profile and was hiding in Libya's southern desert areas.

Before getting to the last brother, three additional terrorist cousins were like brothers to the seven Shaalali brothers. First was Walid Juma Masoud al-Shaalali, a former Abu Salim prisoner like Omar. After being released from Abu Salim prison, he fled to Iraq and joined AQI with his brothers Khaled Juma Masoud al-Shaalali and Mohammad Juma Masoud al-Shaalali. Walid may have been detained by the LNA in February 2019 during counterterrorism operations in Darnah against the Darnah Mujahideen Shura Council (DMSC). His other two brothers were reportedly killed in Iraq.

Omar's last brother, Adil, was a hero in AQI circles and, like his brothers, fought in Iraq. It was Omar who had reportedly facilitated brother Adil to Iraq via Egypt, then Jordan, and then to Damascus, Syria. In 2009, the U.S. Military arrested Adil. He was turned over to Iraqi Military custody and detained. Besides the Shaalali's brothers' perceived grievances against the U.S. for all the family members who died in Iraq, one of the key motivations for the Shaalali brothers to participate in the attack on the U.S. Consulate in Benghazi was due to the fact it was a kidnapping operation. They hoped their brother Adil could be released as part of a prisoner exchange for Ambassador Stevens.

In the months following the attacks, AAS-B had printed out posters that included Adil's image and led protests in both Benghazi and Tripoli on behalf of Libyans imprisoned in Iraq, including large demonstrations on December 16th, 2012. In 2013, the Shaalali brothers were set off by an announcement regarding Adil, who was on death row, awaiting his sentence of execution. In revenge, in November 2013, Omar and his brothers in AAS-D, Fathi and Mohammad, kidnapped and executed an Iraqi professor Dr. Hamid Khalaf Hassan al-Saadi, from Darnah, Libya. They filmed the execution and released the video publicly. As no successful negotiations for a prisoner transfer ever materialized, on August 21st,

2016, Adil was executed in Iraq on terror charges. Specifically, he had intended to be an ISIS suicide bomber looking to seek revenge against the U.S. Military for the death of close associate and former AQI Leader Abu Musab al-Zarqawi. The Government of Iraq returned his body to Libya, arriving at the al-Baydah International Airport on September 8th, 2016.

On November 26th, 2015, Omar and several other terrorists were killed by the LNA in the Battle of al-Aziyat. Al-Aziyat was south of the city of Darnah and on the al-Kharouba al-Tamimi Road. The additional terrorists included: Firas Abdullah al-Amami, Mohammed bin Khayal Arif al-Mansouri, Sharif al-Mansouri, Anis al-Amami, Abdul Rahman al-Haddad, Faraj al-Zobik, Monsef al-Khuram, Hamza al-Jazoi, Abdullah al-Mansoori, and Abdul Hamid al-Shaeri who was with Omar at the Benghazi Consulate attack.

Chapter 6

Enemies: Power Players—Leadership and Commanders

"Power Players" encompasses the senior leaders who brought the U.S. Consulate in Benghazi (and CIA Annex) attacks to fruition on behalf of then al-Qa'ida Leader Dr. Ayman al-Zawahiri.

Included herein are: (6) Harith al-Nadhari, AQAP Senior Sharia Official who issued the fatwa to justify the attack under Islamic principles; (7) Yahya Abu al-Hammam, Deputy AQIM Leader who planned the kidnapping operation with Abdel Malek Droukdal and Mokhtar Belmokhtar; (8) Mohammed al-Gharabi, Rafallah al-Sahati Brigade Leader, who was the most senior al-Qa'ida Commander based in Benghazi in 2012; (9) Ismail al-Sallabi, 17 February Martyrs Brigade and Rafallah al-Sahati Brigade Founder, second most senior al-Qa'ida Commander based in Benghazi in 2012; (10) Mohammad al-Zahawi, Ansar al-Sharia-Benghazi (AAS-B) Leader who planned both the 2012 U.S. Consulate attack and the 2013 In Amenas, Algeria attack with AQIM's Mokhtar Belmokhtar; (11) Hashem Bousidra, AQIM's Mali Group, who was the Mali Group Commander at the U.S. Consulate attack; (12) Boubaker al-Hakim, al-Qa'ida Commander who operated al-Qa'ida's training camp infrastructure at its base in Benghazi in 2012; (13) Abu Bara al-Jazairi, AQIM's al-Mulathameen Battalion who lead Mokhtar Belmokhtar's personal Battalion attackers at the U.S. Consulate attack; (14) Mohammad Jamal, al-Qa'ida in Egypt (AQE) Leader who supplied attackers from Egypt for the U.S. Consulate attack; (15) Murjan Salim, AQE Senior Leader who was involved in promoting al-Qa'ida kidnapping for prisoner swap efforts; (16) Salem Darby, Abu Salim Martyrs Brigade Leader, who supplied attackers from Darnah, Libya for the U.S. Consulate attack; (17) Sufyan bin Qumo, Ansar al-Sharia-Darnah Leader who also supplied attackers from Darnah, Libya for the U.S.

Consulate attack; (18) Boka al-Oraibi, Libya Shield Two Leader who supported the attack on the CIA Annex on September 12, 2012; and (19) Abu Khalid al-Madani, AAS-B's senior Sharia official who provided religious top cover for the group's involvement in the U.S. Consulate attack.

(6) Harith al-Nadhari, full name Harith bin Ghazi al-Nadhari, from Yemen. Harith was a prominent leader in AQAP and was a senior sharia official or religious legal authority. He was also a former member of the Muslim Brotherhood in Yemen's Party for Reform. Harith issued the fatwa or religious ruling before the September 2012 attack on the U.S. Consulate in Benghazi. This is what justified the attack under Islam for al-Qa'ida to be able to partake.

Harith and Nasir al-Wuhayshi, the leader of AQAP, were incredibly loyal to then al-Qa'ida Leader Dr. Ayman al-Zawahiri, who directed the attacks on the U.S. Consulate. Wuhayshi, personally financed a large number of Egyptian attackers affiliated with AQE to prepare, train, and participate in the attacks. In 2013, Zawahiri chose Wuhayshi to be the Deputy Leader of al-Qa'ida. Until his death in 2015, it was believed that Zawahiri might request that Wuhayshi be made the next leader of al-

Qa'ida upon his passing.

For added context, throughout 2012 and 2013, several Yemeni preachers and religious legislators visited Benghazi. On their visits, they held sermons for the various extremist elements in the city. It led to a point where a banner of Osama bin Laden was raised in the city's streets. Two of our Benghazi attackers, Mahmoud al-Awami and Yousef al-Awami, hosted these extremist Yemeni scholars. It was the Awami brothers who also hosted a number of Egyptian Benghazi attackers, the ones that had been funded by AQAP to participate in the attacks. On September 10th, 2012, the Awami brothers used their house as a staging base for the following day's attack on the Consulate.

In July 2014, Nadhari, along with other AQAP seniors, issued a statement defending the leadership of Zawahiri and the core al-Qa'ida ideologies within the movement. The statement was to push back on ISIS as the group was forming its own narratives pitting their operational momentum and successes against those of al-Qa'ida. Nadhari was a known ISIS critic and attempted to use religious justifications to undermine the group.

On January 9th, 2015, Harith delivered a speech regarding the Charlie Hebdo shooting. The attack had occurred just two days prior, killing 12 persons in Paris, France. Paris attacker Cherif Kouachi had admitted before his death that he was directed and financed by AQAP. A senior AQAP official, Nasser bin Ali al-Ansi, claimed credit for the attack and noted it was done following the wishes of Zawahiri. At least one Benghazi attacker was also linked to the Charlie Hebdo shooting, Boubaker al-Hakim. Boubaker, in 2012, led all of al-Qa'ida terrorist training at its Benghazi base in Ganfouda.

On January 31st, 2015, Harith was killed in Shabwa province, Yemen, with three other AQAP terrorists due to a targeted U.S. airstrike. The three additional terrorists included Abdul Samih Nasser al-Hada, Azzam al-Hadrami, and Awad Bafarj.

(7) **Yahya Abu al-Hammam,** real name Djamel Okacha variant Jamal Okasha, from Reghaia, Algeria. At the time of the U.S. Consulate attacks in 2012, Yayha was the Leader of West Africa and the Sahel for AQIM, referred to as the ninth region in al-Qa'ida. He was second-in-command to the Leader of AQIM, Abdel Malek Droukdal. Yahya was in Algeria's first generation of terrorists and was influenced by Hassan Khattab. In the 1990s, Yayha was a Salafist Group for Preaching and Combat (GSPC) member. Yahya was a former prisoner in Algeria for eighteen months as part of the crimes surrounding the "Black Decade," which occurred from 1991 to 2002 when over 200,000 civilians died in the country. In 2004, he first started fighting with Benghazi attacks Mastermind MBM when he joined him to fight in northern Mali and Mauritania.

Yahya planned several terrorist operations and kidnappings. On June 4th, 2005, Yahya with MBM planned the attack at the army barracks in Lemgheity, Mauritania, close to the Algerian and Malian borders. On June 23rd, 2009, Yayha was involved in the murder of Christopher Ervin Leggett, an American evangelist, who was killed in broad daylight in downtown Nouakchott, Mauritania. On August 9th, 2009, he was involved in deploying the first Mauritanian suicide bomber, Ahmed Vih al-Barka, to attack the French Embassy in Nouakchott. On September 11th, 2012, Yayha with MBM was involved in planning the U.S. Consulate attack in Benghazi. On January 16th, 2013, Yayha with MBM was involved in planning the In Amenas oil facility attack in Algeria. This attack used at least a dozen of the same terrorists at the Benghazi attacks.

In March 2017, Yayha announced the establishment of Jamaat Nusrat al-Islam wal Muslimeen (JNIM), also known as the Group for Support of Islam and Muslims, comprised of several key terrorist groups in the region. Those groups included the Sahara branch of AQIM, al-Mourabitoun joined with Ansar Dine, and the Macina Liberation Front (MLF). After the October 2018 arrest of Egyptian terrorist Bahaa Ali Ali Abu al-Maati, he noted the Yayha was providing funding to al-Mourabitoun in Darnah. Specifically, he noted that Mohammad Fathi Karim with alias Abu Malik returned to Darnah from a trip to Mali with 30,000 Euros provided by Yahya. Yayha's close associate, MBM, had been one of the co-founders of al-Mourabitoun in Darnah. Benghazi attacker Marei Zoghbi, historically known for his participation in al-Qa'ida's Milan Cell, was also one of the co-founders.

On November 29th, 2018, Yahya was unsuccessfully targeted in a U.S. airstrike against a vehicle convoy near Al Uwaynat, Libya, which killed 11 terrorists. Then on February 21st, 2019, the French Armed Forces, during Operation Barkhane, killed Yayha in Elakla, northern Mali.

(8) Mohammed al-Gharabi, full name Mohammed Mahmoud Saleh al-Gharabi and alias Sheikh Mohammed from Benghazi. Gharabi was a member of core al-Qa'ida and traveled to Afghanistan during the Soviet-Afghan War in approximately the late 1980s; and fought with and was a close associate of Osama bin Laden. After returning from Afghanistan, he was detained for his membership in al-Qa'ida and served in Abu Salim prison. After the Abu Salim prison massacre in 1996, he joined the newly established al-Qa'ida-affiliated Martyrs Group, which followed guidance from Bin Laden.

Just before the revolution, Gharabi was released from prison. In Benghazi, Gharabi was al-Qa'ida's most senior operational commander, and when foreign fighters from al-Qa'ida started arriving in 2011 and in the years that followed, he was one of the senior al-Qa'ida operatives in Benghazi that they would report to.

He and Ismail al-Sallabi, Ashraf bin Ismail, Fawzi Bu Khatif, Raf Allah al-Sahati, and Mustafa al-Sahati also created the 17 February in 2011. What was not evident from the outside was that Gharabi was treated as the senior official in 17 February as deference to his al-Qa'ida role. Gharabi also led efforts to create the Rafallah al-Sahati Brigade, an operational cell, within 17 February in honor of Raf Allah, who was killed fighting in the revolution. Rafallah al-Sahati Brigade was the most extreme part of 17 February.

Gharabi, just like Mohammad al-Zahawi, provided attackers to MBM (who also fought in Afghanistan) for the U.S. Consulate in Benghazi attacks on September 11th, 2012. He was much more skilled in having a hidden hand, but his involvement is evident now that the attackers have been identified. Gharabi, along with Annex attack Mastermind Wissam bin Humaid, also told the State Department's Principal Officer at the U.S.

Consulate on September 9th, 2012, that they would no longer provide security for Americans in the city—forecasting what was to come. A core al-Qa'ida member having a seat at the table with Americans was just one of many Intelligence Community failures in Benghazi.

As a reminder, fellow 17 February co-founder Fawzi bu Khatif also told our COB "Bob" in the first phone call he received from him the night of the attacks that he would NOT be sending any support. Fawzi got out of the attacks squeaky clean due to his relationship at the time with Bob. Bob personally relied on Fawzi for security in Benghazi—despite pushback from us in the CIA staff. However, a lack of collection does not make you innocent. He was complicit and refused to render aid but still went on to serve in the foreign service with the Government of Libya as the Libyan Ambassador to Uganda.

As the attacks were ongoing on September 11th, Gharabi offered for the CIA personnel to come over and hunker down for safety at the Rafallah al-Sahati Brigade headquarters. Even though the CIA was unaware of Gharabi's position within al-Qa'ida, it was an impossible ask anyway, as, at headquarters, the group had been holding seven Iranians hostage who they detained in July 2012. While the press reported that the Iranians had been working for the Red Crescent, most were Intelligence Officers from Iran's Islamic Revolutionary Guard Corps Quds Force (IRGC-QF).

This detainment actually accounted for all of the Intelligence Officers the Iranians had posted in east Libya for IRGC-QF. This obviously made Iran a moot point regarding any involvement in the Benghazi attacks, so we can shelve that conspiracy theory here and now. Although, the incident did offer another example of a successful kidnapping operation to free al-Qa'ida detainees, as the IRGC-QF officers were swapped for senior Libyan AQI detainees held in Iraqi custody. A total of five AQI members were released in exchange. At the time, there was also an effort to include Adel al-Shalaali, the brother of Benghazi attackers Omar and Mansour. However, their brother wasn't on the final list that the Iraqi Government agreed to and released in October 2012.

Being able to capture the IRGC-QF and essentially get away with it, showed the type of power player Gharabi was in Libya. Besides his role

Enemies: Power Players—Leadership and Commanders

with al-Qa'ida, the strength of his militia battalions alone gave Gharabi the power to influence political decisions. Besides having a relationship with U.S. officials in Benghazi, Gharabi was supported by the Government of Libya in Tripoli. Leadership in Tripoli enabled Gharabi by financially supporting him in his role as a government "Advisor"—even as he controlled the largest number of terrorist training camps which belonged to al-Qa'ida in the vicinity of Benghazi. He also facilitated foreign fighters to Libya on behalf of al-Qa'ida. When he finally fled Benghazi, he relocated to Tripoli, where he resided as of 2022. He also traveled freely to Istanbul, Turkey.

Ismail al-Sallabi
Deputy Commander of al-Qa'ida in Benghazi
Founder of 17 February
STATUS: AT-LARGE, AS OF 2022, LIVES IN TURKEY, FREQUENTS LIBYA

PERSONAL INFO
Full Name: Ismail Mohammad al-Sallabi
Affiliation: Al-Qa'ida, Libyan Islamic Fighting Group (LIFG), 17 February
Nationality: Libyan

BIOGRAPHY
In the 1980s, fought in Afghanistan against Soviet Forces; He and Mohammad al-Gharabi were the most senior al-Qa'ida operatives in Benghazi after the Libyan Revolution

In 1997, arrested for harboring terrorists; Sent to Abu Salim; Released

By 2004, had fled to Europe, arrested for suspected involvement in the July 7, 2005 London bombings

On September 11, 2012, over 30 terrorists who had been members of 17 February or Rafallah attacked the U.S. Consulate; Further, 17 February refused to send back up forces to assist

In May 2017, carried out the Brak al-Shati massacre, killed over 140

(9) Ismail al-Sallabi, full name Ismail Mohammad al-Sallabi from Misrata, Libya. Ismail was the Founder of the 17 February Martyrs Brigade and the Rafallah al-Sahati Brigade with Gharabi. In the weeks that followed the attacks, both militias signed up under the umbrella organization of Libya Shield 7 and later joined the Benghazi Shura Revolutionary Council (BSRC).

In the 1980s, Sallabi fought in Afghanistan against Soviet Forces. Sallabi became a senior leader in the LIFG. In 1997, he was arrested for harboring terrorists as he was essentially running an underground railroad for terrorists in and out of Libya in support of both al-Qa'ida and the LIFG. He was sent to Abu Salim and was released by early 2004. His release was partly due to assistance from the Qatari Government and thanks to efforts by his brother, Qatari-based cleric Ali Mohamed Mohammed al-Sallabi, known better as Ali al-Sallabi. In 2011, during the Libyan revolution, Ali helped funnel Qatari arms shipments to the revolutionaries in the fight against Gaddafi.

After Sallabi's prison release in 2004, he was constantly monitored by Gaddafi's internal security services. He decided to flee to the United Kingdom to get out of the eye of Gaddafi's intelligence teams. In July 2005, British authorities arrested him for suspected involvement in the July 7th, 2005, London bombings, the series of four coordinated suicide attacks that killed 52 citizens and injured 784. After his release, Sallabi returned to Libya. In 2011, Sallabi fought in the Libyan revolution with his group 17 February.

On September 11th, 2012, many current and former members of both 17 February and Rafallah al-Sahati were involved in the attacks. At the time, nothing major happened in Benghazi without Sallabi, Mohammad al-Gharabi, Wissam bin Humaid and Mohammad al-Zahawi being aware, giving consent, and likely being directly involved. Sallabi, like Wissam, was a master at keeping a hidden hand even though he and Gharabi were the two most senior historic al-Qa'ida members in Benghazi. Sallabi's three brothers-in-law were attackers at the U.S. Consulate, including Talal, Faisal, and Hudhayda bin Hariz.

In 2014, Sallabi co-founded the al-Qa'ida-affiliated BRSC and, in 2016, was one of the co-founders of the Benghazi Defense Brigades. Both groups led several high-profile terrorist attacks against the LNA. For example, in May 2017, he carried out the Brak al-Shati massacre against the LNA, killing over 140. Sallabi attacked the military base with Benghazi attackers Amin Kelfa, Faraj Shakku, and Ahmed al-Tajouri. After this attack, Sallabi moved over to fight the LNA in Libya's capital of Tripoli and, as of late 2017, was the commander of the Tripoli Brigade.

Over the years, Sallabi had been involved in terrorist attacks, targeted assassinations, money laundering, foreign fighter pipelines, criminal enterprises, and weapons trafficking. Still, he was able to carry out all these activities unabated. In our listing of U.S. Consulate in Benghazi attackers, over 30 of them had either been a member of the 17 February Martyrs Brigade or the Rafallah al-Sahati Brigade, the two key organizations founded by Sallabi and Gharabi in 2011. Key al-Qa'ida terrorist leaders like Sallabi and Gharabi should have bounties on their heads, they should not travel freely, and they should be brought to justice. Sallabi, most recently was involved in supporting terrorism efforts from Turkey into Libya. Terrorists like him only become more emboldened when their actions have no repercussions. As of 2022, Sallabi was at-large living in Turkey with his brother Osama, but frequented Misrata, Libya. His brother Ali al-Sallabi, was located in Qatar. And the brother reported to be most extreme, Khaled al-Sallabi, lived in the United Kingdom.

(10) Mohammad al-Zahawi, full name Mohammad Ali al-Zahawi with alias Abu Musab from the al-Majouri neighborhood in Benghazi. In September 2012, Zahawi was the Leader of Ansar al-Sharia-Benghazi (AAS-B) which he was a co-founder of in February 2012. Zahawi formed AAS-B with a number of terrorist associates, many of whom eventually participated in the 2012 Consulate attacks. Zahawi maintained a close relationship with U.S. Consulate Mastermind, Mokhtar Belmokhtar (MBM), and Zahawi assisted him in planning two key attacks, the U.S. Consulate in Benghazi attack in September 2012 and the In Amenas, Algeria attack in January 2013.

In the mid-1990s, after Osama bin Laden departed Afghanistan for Sudan, Zahawi went to Sudan to meet up with him. Even years before the 9/11 attacks, Bin Laden was lauded as a hero in terrorist circles as he was a key leader in the jihad against the Russians to expel them from Afghanistan. Zahawi was detained by Sudanese authorities and extradited to Libya, where he was sent to Abu Salim prison. Like many attackers, Zahawi was freed from prison during the Libyan revolution in 2011. After leaving prison, Zahawi went straight to Misrata, Libya, to fight against Gaddafi forces with terrorists Ahmed al-Mushaiti, Mohammad al-Ferjani, Ali al-Karshini, and Mohammad al-Manfi. Mushaiti, Karshini, and Manfi all participated with Zahawi in the September 11th, 2012, attacks on the U.S. Consulate.

At the end of the battles in Misrata, Zahawi returned home to Benghazi for the first time in many years. He initially joined the Rafallah al-Sahati Brigade, which at the time was a cell under the 17 February led by Fawzi bu Khatif and Ismail al-Sallabi. This militia included terrorists recruited from across several historic regional terrorist organizations, including al-Qa'ida, the LIFG, Ansar Allah, and the Martyrs Group. By 2012, Zahawi and associates had the idea to combine all the most extreme organizations under one umbrella, which became AAS-B. Zahawi believed bringing like-minded extremists together would help create a power base. He was concerned that several groups (like 17 February, Rafallah al-Sahati, and the Zintan Martyrs Brigade) were trying to interact with and, in Zahawi's opinion, "appease" the international community. A conglomerate could oppose western influence and counter efforts by these groups to form security cooperation agreements with the international community.

AAS-B officially announced its formation in the summer of 2012 in preparation for the group's intent to declare the city an Islamic emirate affiliated with al-Qa'ida. The group held its first annual Islamist conference from June 7th to 8th in the Liberation Square in Benghazi. AQIM's MBM was in attendance along with a number of regional terrorist groups, and at least 15 Libyan militia organizations. The idea came from the successful conference hosted by Ansar al-Sharia-Tunisia (AAS-T) in 2011 and the subsequent conference held in 2012, which had over 10,000 attendees. After the Benghazi conference, AAS-B umbrella members became more involved in terrorist-related activities. For example, on June 18th, 2012, when a group of 20 armed AAS-B members stormed the Tunisian Consulate in Benghazi and raised their black flag over the Consulate.

Zahawi was targeted in an LNA airstrike and succumbed to his injuries on January 23rd, 2015. Nasser al-Tarshani, with alias Khalid al-Madani, succeeded Zahawi. After his death, Zahawi was eulogized personally by then al-Qa'ida Leader Zawahiri; by Senior AQAP Sharia Official Harith al-Nadhari; by AQIM leadership; and by Yahya Abu al-Hammam's JNIM an al-Qa'ida coalition linked to both AQIM and MBM's al-Mourabitoun Battalion. When Nasser al-Tarshani took over, he did not command the same presence as Zahawi. This affected his ability to keep AAS-B together, and members of the group, began to defect to ISIS. Further, in early 2015, the LNA also took over the Port of Benghazi, AAS-B's key base of operations. In May of 2017, the group disbanded.

(11) Hashem Bousidra, full name Hashim Hussein Abdul Jawad Bousidra with aliases Abi Ahmed and Khubayb variant Khabib al-Darnawi from Darnah. Hashem is the brother of former Libyan General National Congress member Mohammad Bousidra. In 2003, Hashem joined local extremist groups and fled to Algeria in 2005, where he joined AQIM and fought in Algeria and Mali. He was known for smuggling weapons and terrorist fighters within the Sahel for AQIM. In 2011, he returned from Algeria and fought in the Libyan revolution.

By 2012, he was one of the senior leaders within the Mali Group with Mahdi Dango with alias Abu al-Barakat and Jaafar Azzouz. The Mali Group was a cell within AQIM. On September 11th, 2012, he attacked the U.S. Consulate in Benghazi with Mali Group members, including long-time associates Mahdi Dango, Abu Hamza al-Tabawi, Abdul Hamid al-Shaeri, and Mansour al-Shalaali. Mansour was the brother of Omar al-Shalaali, the AQIM Leader for Libya at the time, and he was the al-Qa'ida Commander that led the attacks on the ground at the Consulate.

After the attacks, Hashem linked up with Abu Salim Martyrs Brigade (ASMB) in Darnah. When ISIS came into Darnah, he pledged allegiance to the group in 2014. On February 12th, 2015, he, along with Benghazi attackers Mahdi Dango and Jaafar Azzouz, and terrorist Walid al-Ferjani were involved in the kidnapping of 21 Egyptian Coptic Christians and then carried out the massacre against them on February 15th, 2015.

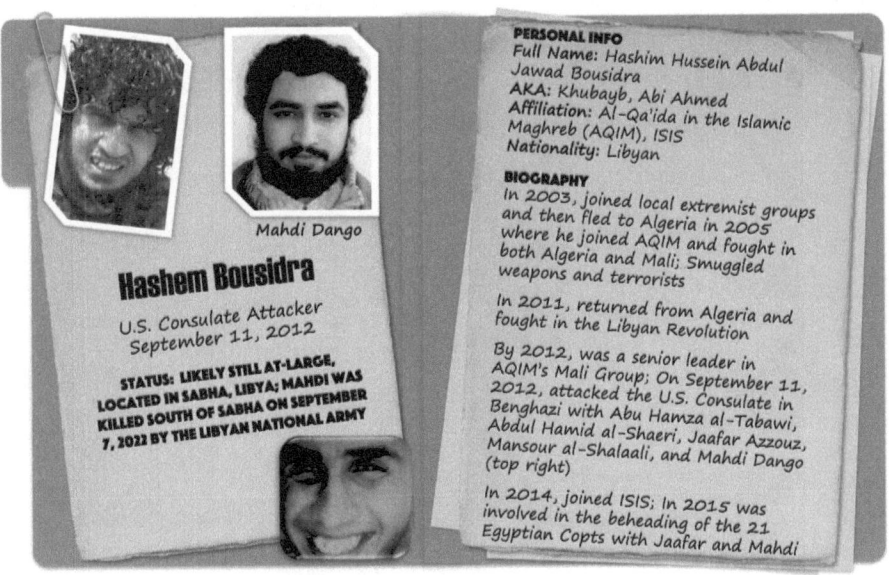

Hashem relocated to Sirte, Libya and he was the ISIS Border Commander leading the Foreign Transport Administration. As Border Commander, he was responsible for the movement of terrorists in and out of Libya, including for terrorist training. In 2016, Hashem fled Sirte just before the kick-off of Operation Al-Bunyan al-Marsous, an operation launched by Libya's Government of the National Accord (GNA) against ISIS.

On September 19th, 2019, when close associate and Benghazi attacker Abu Hamza al-Tabawi was killed in an airstrike, Hashem was reportedly in the vicinity, and it was rumored the strike may have targeted him. As of 2022, he was at-large and located in Sabha, Libya. On September

7th, 2022, the LNA killed fellow Mali Group senior leader and fellow Benghazi attacker Mahdi Dango just south of Sabha.

(12) Boubaker al-Hakim, full name Boubaker ben Habib ben al-Hakim variant Abu Bakr al-Habib 'Abd al-Hakim with alias Abu Muqatil al-Tunisi from France. Boubaker was a former member of al-Qa'ida in Iraq (AQI) and was also closely associated with Ansar al-Sharia-Tunisia (AAS-T) Leader Seifallah Ben Hassine. In 2012, he was running al-Qa'ida's primary terrorist training camp in the Ganfouda neighborhood in Benghazi, which he opened just after the Libyan revolution in early 2012. Ganfouda was where al-Qa'ida established its primary base in Libya, starting just after the revolution in late 2011 until its defeat by the LNA in January 2017.

The U.S. invasion of Iraq, in 2002, is what catapulted Boubaker into terrorism. He traveled to Syria to support jihad, returned to France briefly, and then traveled to Iraq. In Iraq, he went to fight with Abu Musab al-Zarqawi and his group at the time called Jama'at al-Tawhid wa'l Jihad (JTJ). Soon after, Zarqawi rolled JTJ into AQI. Boubaker fought in Iraq until fleeing to Syria in May 2004 after fighting the U.S.

Military in Operation Vigilant Resolve or the First Battle of Fallujah; the counterterrorism operation ran from April 4th, 2004, to May 1st, 2004, in Fallujah, Iraq.

The operation was kicked off, after Boubaker and other members of JTJ, on March 31st, 2004, ambushed four of our Blackwater brothers, Scott Helvenston, Jerry Zovko, Wesley Batalona, and Mike Teague. His network was reportedly involved on the same day in an improvised explosive device (IED) attack that killed an additional five Americans serving our country near Habbaniya, Iraq. Those we lost included 1st Lt. Doyle M. Hufstedler, Army Spc. Sean R. Mitchell, Spc. Michael G. Karr Jr., Pfc. Cleston C. Raney, and Pvt. Brandon L. Davis. The five hailed from the Army's 1st Engineer Battalion, 1st Brigade, 1st Infantry Division out of Fort Riley, Kansas.

While in Syria, Boubaker was captured by Syrian Intelligence and extradited to France on terrorism charges. In 2008, he was sentenced to 8 years in prison for running the Buttes-Chaumont terror network, a network to recruit fighters from Europe and send them to fight in Iraq. Somehow, Boubaker was released from French custody, serving less than three years, and showed back up in Tunisia in 2011 during the Arab Spring. After being involved in events in Tunisia, he relocated to Benghazi. In early 2012, he set up a terrorist training center for al-Qa'ida with Tunisian terrorist Al-Bashir bin Yahya al-Zanqah alias Bin Saber in an old Libyan Army base in Ganfouda. Boubaker would train foreign fighters for al-Qa'ida, who then would deploy from Benghazi to conflict areas including Iraq, Syria, and Somalia. Besides Ganfouda being the primary deployment base for al-Qa'ida in the world at the time, al-Qa'ida was also using this location near the port to ship out weapons and ammunition, taking advantage of former Gaddafi weapons stockpiles and the weapons pipelines moving freely through Libya at the time.

On September 11th, 2012, Boubaker attacked the U.S. Consulate in Benghazi. Like most attackers, Boubaker believed the U.S. would bomb al-Qa'ida's base in Ganfouda and immediately fled to Syria with many of the attackers. It was completely unknown to the CIA at the time that al-Qa'ida had established a base in the city. We actually only had one counterterrorism case officer even serving in Benghazi in 2012. The reality was that CIA was too preoccupied collecting on political reporting in eastern Libya at the direction of COB Bob, who was laser-focused on a political issue (that never mattered at the end of the day) which was termed "Federalism."

In August 2012, Boon and another GRS officer went to a suspected terrorist camp in Benghazi to confirm reporting that it was training foreign fighters. After reaching the camp, the two decided to climb the walls of the camp to get a better look. When looking in, they saw an individual who resembled Boubaker, and reported back that al-Qa'ida fighters were located at the camp, not Libyan militia members. Both were then reprimanded by Bob at the CIA Annex for collecting on the camp and told they could never return to the area as he did not want the terrrorists to feel they were being surveilled. Further, CIA was also forbidden then from doing any operations in the area. It was unclear if this potential sighting of Boubaker was ever reported in CIA channels.

After arriving in Syria after the attacks, Boubaker reconnected with former AQI regional associates and fought with them until they pledged allegiance to ISIS. He had consistently traveled back and forth between Libya, Syria, and Tunisia. In 2013, with Benghazi attacker Ali Ouni al-Harzi, Boubaker was involved in two high-profile assassinations in Tunisia. The February 6th, 2013, assassination of Tunisian opposition party leader Chokri Belaid; and then the July 25th, 2013 assassination of another Tunisian opposition leader, Mohamed Brahmi.

Boubaker returned to Libya and settled near the Tunisian border. In 2015, he was reported to have played a role in two international terrorist attacks. The first attack was the January 7th, 2015, attack on the offices of the French satirical weekly newspaper Charlie Hebdo in Paris, France, which killed 12 people and injured 11 others. And the second set of attacks was "Bloody Friday," which occurred over two days, from June 25th to June 26th, 2015, with synchronized terrorist attacks occurring in France, Kuwait, Syria, Somalia, and Tunisia, killing over 400 persons. After these attacks, counterterrorism operations heated up in the Iraq and Syria regions against ISIS and other militants affiliated with groups like al-Qa'ida.

This is where a conflict in his biography comes into play. His wife, whom he met through Facebook and married in 2012, and who traveled with him to Libya in 2012, escaped from him in Libya in 2016. At the time, she returned to Tunisia. According to her timeline and others on the ground who saw him in Libya after the strike, he may not have been located in Raqqa, Syria, on November 26th, 2016, when the U.S. reported they had killed him in an airstrike. Separately, in 2017, the Iraqis noted that they were targeting Boubaker with the Syrians.

Due to these discrepancies, and since we could not secure a post-mortem photograph of him, we note that his status is unknown at this time. He is likely deceased due to a lack of reporting on him since.

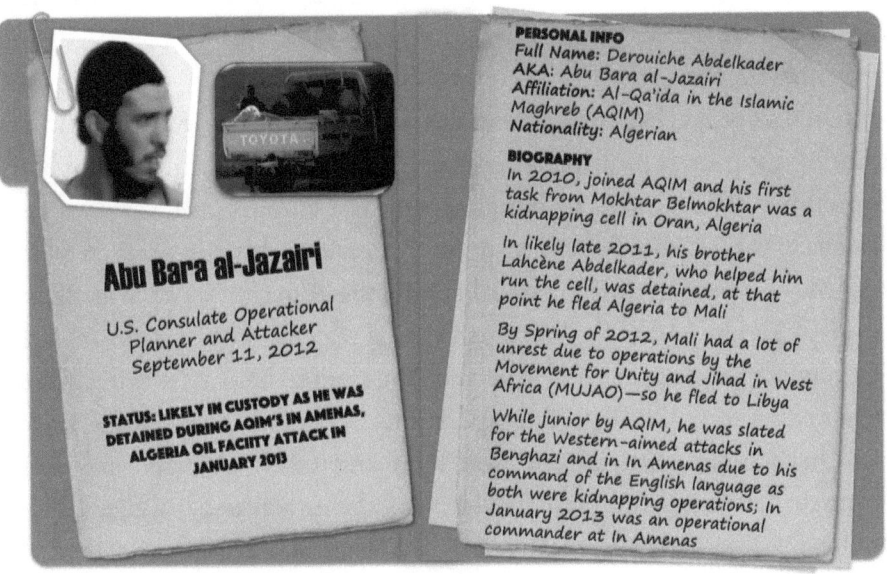

Abu Bara al-Jazairi

U.S. Consulate Operational Planner and Attacker
September 11, 2012

STATUS: LIKELY IN CUSTODY AS HE WAS DETAINED DURING AQIM'S IN AMENAS, ALGERIA OIL FACLITY ATTACK IN JANUARY 2013

PERSONAL INFO
Full Name: Derouiche Abdelkader
AKA: Abu Bara al-Jazairi
Affiliation: Al-Qa'ida in the Islamic Maghreb (AQIM)
Nationality: Algerian

BIOGRAPHY
In 2010, joined AQIM and his first task from Mokhtar Belmokhtar was a kidnapping cell in Oran, Algeria.

In likely late 2011, his brother Lahcène Abdelkader, who helped him run the cell, was detained, at that point he fled Algeria to Mali

By Spring of 2012, Mali had a lot of unrest due to operations by the Movement for Unity and Jihad in West Africa (MUJAO)—so he fled to Libya

While junior by AQIM, he was slated for the Western-aimed attacks in Benghazi and in In Amenas due to his command of the English language as both were kidnapping operations; In January 2013 was an operational commander at In Amenas

(13) Abu Bara al-Jazairi, with real name Derouiche Abdelkader from the Wilayat of Tiaret, in the town of Takhmart, Algeria. In 2010, Abu Bara was a businessman who joined AQIM after MBM took over control of his town. His first task from MBM was to set up a kidnapping for ransom cell in Oran, Algeria. The operation was successful until Abu Bara's brother, Lahcène Abdelkader, who assisted him in operating the cell, was detained by Algerian authorities. At this point, Abu Bara fled Algeria for Mali. However, Mali was no longer hospitable by the spring of 2012 as the Movement for Unity and Jihad in West Africa (MUJAO) was taking over key cities in Mali. MBM would later, in 2013, merge his Battalion with MUJAO. Abu Bara thought Libya would be the safest place to relocate to as it was becoming a refuge for terrorists from across the region, including foreign fighters returning to Libya.

As noted, AQIM was laser focused on expanding its presence in Libya in 2012 as it was an operational priority for the group's leader Droukdal. Abu Bara traveled with MBM as he worked to establish relationships throughout Libya, especially in eastern Libya. As a key operational planner in MBM's Battalion, Abu Bara was trusted to assist in fostering relationships in Ajdabiya. He also maintained communications and relationships with various terrorist groups and militias in Libya throughout 2012.

Further, Abu Bara was responsible for ensuring that terrorists affiliated with AQIM were able to enroll in training camps in Libya. This training supported multiple attacks planned by AQIM, most notably the September 2012 attacks in Benghazi and the January 2013 attacks in In Amenas. While Abu Bara was junior concerning his time served within AQIM, as other terrorists dated back to the Algerian Civil War, he was hand-picked by MBM for Western-focused attacks due to his command of the English language. MBM expected Abu Bara to help assist with western detainees at several kidnapping operations he had in the planning stages.

On January 16th, 2013, Abu Bara and fellow AQIM operational commander Abdul Rahman al-Nigeri led the attack and hostage standoff at the Tigantourine gas facility near In Amenas, Algeria. Abu Bara and 11 attackers were present at both the In Amenas and Benghazi attacks, including 2 Canadian AQIM terrorists, Xris Katsiroubus and Ali Medlej. On January 19th, Abu Bara was detained at the end of the crisis; his fellow operational commander Abdul Rahman and both Canadians, Xris and Ali, were killed.

(14) Mohammad Jamal, full name Mohammed Jamal Abd al-Rahim Ahmed al-Kashif with alias Abu Ahmad, from the Shubra al-Khaimah area of Cairo, Egypt. The Leader of al-Qa'ida in Egypt (AQE), at the time of the U.S. Consulate in Benghazi attacks on September 11th, 2012. Jamal was reportedly in Benghazi for the attacks and trained a number of AQE terrorists that participated. His AQE terrorists were funded by Nasir al-Wuhayshi, the Leader of AQAP.

Jamal was a close confidant to former al-Qa'ida Leader Dr. Ayman al-Zawahiri since the 1980s when Jamal traveled to Afghanistan to serve core al-Qa'ida as an explosives expert. He trained al-Qa'ida trainees at camps in both Pakistan and Afghanistan. At one point, he even served on the security detail for Osama bin Laden. He also was the personal security detail for Zawahiri in the 1990s while both in EIJ.

Upon his return to Egypt from Afghanistan, Jamal was the top military commander and head of the operational wing of EIJ under Zawahiri. As background, Zawahiri led the EIJ and then in 1998 merged it with Osama Bin Laden's al-Qa'ida, the group Zawahiri led until his death in July 2022. Jamal also served in Yemen on behalf of EIJ and it was the reason he had a close relationship to AQAP senior leaders come 2012.

When Jamal returned to Egypt from the Pakistan and Afghanistan border region, likely due to being deported over the killing of a child, he was charged with terrorism under the rule of President Mohamed

Hosni Mubarak. Jamal was sent to the Scorpion Prison, located inside Cairo's Tora Prison Complex. During the Arab Spring in 2011, Egyptian President Mohamed Morsi released Jamal with a presidential pardon.

Jamal formed AQE after his release. He established several terrorist training camps in Egypt and Libya—camps that reportedly trained a number of the 2012 Benghazi attackers. Like with groups in Tunisia and Libya, Jamal could take advantage of the significant release of terrorists from Egyptian prisons. The releases were part of Egyptian Government initiatives in the aftermath of the Arab Spring uprisings. AQAP not only funded AQE's formation, but Jamal had also used the AQAP smuggling network to move local and foreign fighters. He trained suicide bombers at AQE training camps and linked up with terrorists in Europe as AQE wanted to focus on advancing the al-Qa'ida mission against the West.

Due to his direct involvement in the Benghazi attacks, Mohmmad Jamal was re-arrested by Egyptian authorities in November 2012. His computer was exploited, and there was documentation to include letters to Zawahiri in which Jamal asked for assistance to acquire weapons, conduct terrorist training, and establish additional terrorist groups in the Sinai. Jamal's two primary groups formed before his arrest were AQE and an operational cell coined the Nasr City Cell under AQE.

There were two key Jamal associates from the Nasr City Cell that played an important role related to AQE's involvement in the Benghazi attacks—Sheikh Adel Awad Shehato and Tarik Abu al-Azzam with alias Abu Hamza. As background, Shehato was the Leader of the Nasr City Cell, at the time of the attacks. He was previously a senior leader in the EIJ, and spoke on video during the U.S. Embassy in Cairo protests on September 11th, 2012. The following key individuals planned the protests: Mohammad al-Zawahiri, other former EIJ leaders, and the family of Sheikh Omar Abdel-Rahman, better known as the "Blind Sheikh."

After it was reported in the press that AQE was involved in the September 11th, 2012, Benghazi attacks, Shehato felt that Egyptian authorities were closing in on him. Therefore, he attempted to flee to Libya but was caught and detained while crossing the border region.

Separately, Tarik Abu al-Azzam was Jamal's Deputy Leader in AQE at the time of the Benghazi attacks. He had been a former Major in the Egyptian Air Force and, back in the day, had traveled to the U.S. and participated in a Military exchange program. This program was similar to the one Saudi Arabian Military student Mohammed Saeed Alshamrani, who committed the December 6th, 2019, terrorist attack on the Naval Air Station Pensacola in Florida, attended.

Tarik was imprisoned in 2002 for his involvement in a terrorist cell operated by Jundallah. Jundallah was an al-Qa'ida-affiliated group operating out of Gaza. The group was separate from the Iranian Shia Jundallah group and the Karachi-based Sunni Jand Allah group. Two key plots Tarik participated in were first, a plot against U.S. ships transiting the Suez Canal and a second plot to bomb the U.S. Embassy in Cairo. The Cairo Embassy was also a target of AQE in the aftermath of the U.S. Consulate in Benghazi attacks in 2012.

Jamal corresponded directly not just with Zawahiri, but also with the leaders of AQAP and AQIM and terrorists in Palestine. In addition to being close to Zawahiri, Jamal was equally as close to his brother Mohammad al-Zawahiri as they were imprisoned together. Mohammad al-Zawahiri instigated the protests that occurred in Cairo, Egypt, on September 11th, 2012. Mohammad al-Zawahiri was also close to AQE Senior Leader Murjan Salim, who was involved in planning the Benghazi attacks. Egyptian authorities arrested Jamal in a Nasr City Cell raid in October 2012. On October 22nd, 2014, he was sentenced to 25 years in prison.

(15) Murjan Salim, full name Murjan Mustafa Salim el-Gohary and alias Abdul Hakim Hassan from Egypt. Murjan was a core al-Qa'ida member and historically had been an EIJ member. He remained connected to its former Leader Dr. Ayman al-Zawahiri before his death.

Murjan comes from the first generation of al-Qa'ida. In approximately the late 1980s, he fled counterterrorism operations in Egypt as he was targeted for his role in EIJ. He first traveled to Yemen and then traveled to Afghanistan, and also spent time in Iran. In Afghanistan, Murjan administered the publication of al-Qa'ida's magazine called Maalim al-

Jihad. In March 2001, he also oversaw the demolition of the two Buddhas of Bamiyan statues after Taliban founder Mullah Omar ordered their destruction.

Before the Libyan revolution, Murjan was serving in prison in Egypt with several other individuals involved in the planning side of the Benghazi attacks, including Mohammad Jamal, Sheikh Adel Awad Shehato, and Tarik Abu al-Azzam. President Mohamed Morsi released Murjan in 2011. Murjan was a front-facing terrorist and had no qualms about appearing in the media, even though his views were quite extreme. For example, he issued a fatwa to destroy the great pyramids in Egypt.

In the aftermath of the Arab Spring, at least publicly, Murjan acted like a community organizer where he would instigate extremist protests, including in Rabaa al-Adawiya Square. In 2012, he led a demonstration with Safwat Hijazi, Hazem Salah Abu Ismail, Atef Rashid, Omar Rifai Sorour, Mohammad al-Zawahiri (again, brother of Dr. Ayman al-Zawahiri), and thousands of other terrorists. In the large gathering, the terrorists made several requests: 1. Implementation of Sharia law; 2. The release of terrorists imprisoned in Egypt and worldwide with United States, Great Britain, Algeria, and Yemen being mentioned; and 3. A special request for the release of the Blind Sheikh. The Sheikh later died in U.S. prison on February 18th, 2017.

One of the primary objectives of the September 11th, 2012, attacks in Benghazi was to kidnap the Ambassador Stevens to use him for hostage exchanges to free al-Qa'ida terrorists, and their allies captured abroad. Freeing the detainees then would be a symbolic gesture to honor Abu Yahya al-Libi. Murjan was also in a popular video that was making the rounds that focused on releasing detainees from prison. In the video, which the al-Qa'ida-affiliated al-Farouq Media Foundation filmed, several terrorists discuss using fatwas to help release more prisoners. The terrorists in this video with Murjan included Mohammad al-Zawahiri and Sorour again. Of note, Sorour, who was killed in Libya had founded the al-Mourabitoun organization in Darnah in partnership with U.S. Consulate attack Mastermind MBM.

On September 11th, 2012, Murjan and Jamal provided attackers for the Benghazi attack. Murjan, along with Mohammad al-Zawahiri, was also involved in the incident outside the U.S. Embassy in Cairo earlier in the evening. Both were aware an attack was planned for later that evening in Benghazi, by al-Qa'ida, and at the behest of Mohammad's brother. In 2013, Murjan was arrested and sent to Scorpion Prison in Egypt, and his death was announced in August 2015.

As an aside, from 2011 to 2013, the main individual who armed Egyptian terrorists belonging to al-Qa'ida in Libya was Mohammad Abd al Sami Hamida Abd Rabbo. Rabbo supported AQE, the group's Nasr City operational cell, and a separate terrorist cell led by Nabil Mohammad Abd al-Majid al-Maghrebi. Egyptian Intelligence arrested Rabbo in Marsa Matrouh, Egypt, on the border with Libya. As of 2022, Abbo had not yet gone to trial.

(16) Salem Darby[iv], full name Salem al-Barani Darby from Tobruk, Libya, but his family was historically from Darnah. He had a brother, Al-Mahdi, serving in Abu Salim prison since 1995. In 1996, Salem became involved in militancy movements and went into hiding as there was extra surveillance on him by the internal security services. In 2009, Salem noted that his brother had been killed in Abu Salim.

On February 17th, 2011, at the onset of the Libya revolution, Salem, and Abdul Hakim al-Hasadi formed the Abu Salim Martyrs Brigade (ASMB) to honor those who died in Abu Salim. The Brigade was the largest militia operating in Darnah at the time. The Brigade got up to approximately 1000 members by the end of the revolution. The group's first battle was in March at the Battle of Qawarsha in western Benghazi. It is important to note this as several Benghazi attackers fought together in this early battle. In June 2012, Salem took over as sole leader of the group when al-Hasadi decided to run for political office in Darnah. Several LIFG members took the political route after the revolution.

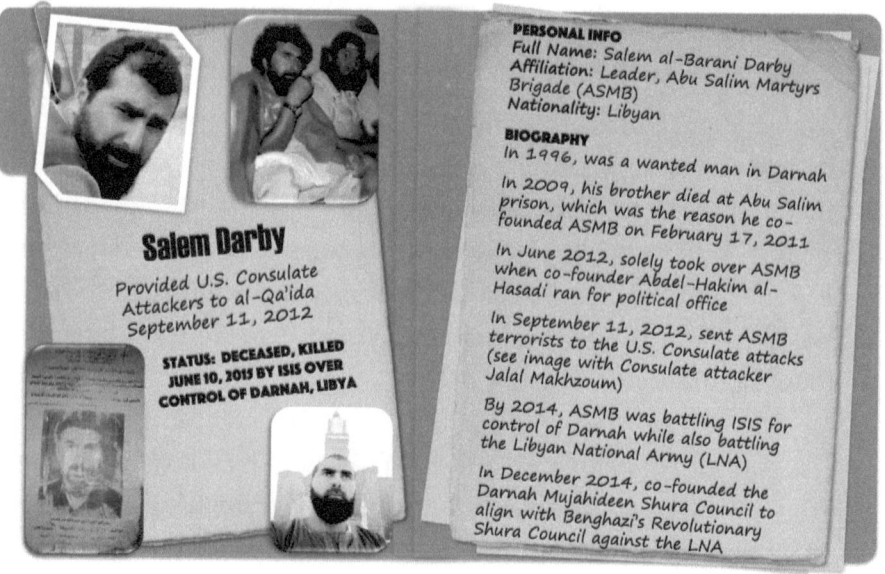

Salem Darby
Provided U.S. Consulate Attackers to al-Qa'ida September 11, 2012

STATUS: DECEASED, KILLED JUNE 10, 2015 BY ISIS OVER CONTROL OF DARNAH, LIBYA

PERSONAL INFO
Full Name: Salem al-Barani Darby
Affiliation: Leader, Abu Salim Martyrs Brigade (ASMB)
Nationality: Libyan

BIOGRAPHY
In 1996, was a wanted man in Darnah

In 2009, his brother died at Abu Salim prison, which was the reason he co-founded ASMB on February 17, 2011

In June 2012, solely took over ASMB when co-founder Abdel-Hakim al-Hasadi ran for political office

In September 11, 2012, sent ASMB terrorists to the U.S. Consulate attacks (see image with Consulate attacker Jalal Makhzoum)

By 2014, ASMB was battling ISIS for control of Darnah while also battling the Libyan National Army (LNA)

In December 2014, co-founded the Darnah Mujahideen Shura Council to align with Benghazi's Revolutionary Shura Council against the LNA

On September 11th, 2012, Salem sent a contingent of ASMB terrorists to participate in the attacks on the U.S. Consulate in Benghazi. It was unconfirmed whether Salem was at the attacks that evening. However, senior operative Abdelkader Abdelsallam Azzouz, al-Hasadi's brother-in-law, was a key ASMB leader on the ground at the Consulate. ASMB member and al-Qa'ida terrorist Amin Kelfa participated in the attacks, as well. Amin was known to operate his own cell within the group suggesting that ASMB had multiple cells participate in the attacks.

Lastly, Omar al-Shalaali who led the attack for al-Qa'ida, had been one of the founders of ASMB.

In 2013, a bomb was placed on Salem's car. The car was parked in front of his house, but it did not injure anyone. Then in 2014, he was ambushed on his farm by ISIS but survived. By 2014, ISIS was in control of much of Darnah after aligning with a local militia, the Shura Council of Islamic Youth (SCIY). For the next year, ASMB and ISIS battled for control of Darnah. In December 2014, Salem founded the al-Qa'ida-affiliated DMSC to align with BRSC against General Haftar's LNA.

On June 10th, 2015, ISIS, in a push to gain control of Darnah, assassinated Salem. He was killed with another senior DMSC militant, Nasir Attiyah al-Akar. Their deaths lead to an outpouring of local militant support in Darnah, giving al-Qa'ida extra momentum to push ISIS out of its area of operations. ISIS was mostly eliminated from Darnah's city center within one month of Salem's death.

Sufyan bin Qumo
Supplied U.S. Consulate Attackers
September 11, 2012
STATUS: UNKNOWN, MIXED REPORTING, POSSIBLY IN LIBYAN NATIONAL ARMY CUSTODY

PERSONAL INFO
Full Name: Sufyan Ibrahim Ahmed Qumo al-Hasadi
AKA: Abu Faris
Affiliation: Al-Qa'ida, Ansar al-Sharia-Darnah (AAS-D), ISIS
Nationality: Libyan

BIOGRAPHY
In November 2002, arrested in Pakistan for being al-Qa'ida; In 2003, was transferred to Guantanamo Bay; In 2007, repatriated to Libya; In 2010, released from Abu Salim prison
In 2012, Leader of AAS-D and sent fighters and trainees to the attacks
In 2014, joined ISIS in Darnah and facilitated the groups local founding
In 2015, detained temporarily by al-Qa'ida's-affiliated Darnah Mujahideen Shura Council (DMSC)
In 2018, was reportedly captured or killed with Egyptian al-Qa'ida leader, Omar Sorour; Other reports place him in hiding in Adana, Turkey

(17) Sufyan bin Qumo, full name Sufyan Ibrahim Ahmed Qumo al-Hasadi with alias Abu Faris from Darnah was a member of al-Qa'ida. In 1991, Qumo's first brush with the law was as a criminal when he was arrested on felony charges related to assault, drugs, and murder. In 1993, he escaped from al-Kuwaifiya prison in Benghazi, and then first fled to Egypt. He reportedly traveled to Saudi Arabia and ended settling in Suda, Sudan, where he first met Osama bin Laden. Qumo, in the 1980s, had driven tanks in Gaddafi's Libyan Army, and this experience led him to be a truck driver for Bin Laden at the Wadi al Aqiq Company. In Sudan, Qumo also established a relationship with the early beginnings of the LIFG, which was established in 1995.

When Qumo was forced out of Sudan in the late 1990s, he first traveled to Britain on a forged passport where several LIFG individuals were located. He then traveled on to support al-Qa'ida efforts in Afghanistan. His cover for his terrorist actions in the Pakistan and Afghanistan border region was with the al-Wafa Charitable Foundation. He would work out of the Foundation and then support terrorists fighting the war against the Northern Alliance in Afghanistan.

When the U.S. invaded Afghanistan, they were able to close in on Qumo with information reportedly provided by Pakistani intelligence, and he was arrested in Pakistan in November 2002. In May 2003, he was transferred to Guantanamo Bay, Cuba. On September 28th, 2007, Qumo was repatriated to Libya and sent to Abu Salim prison. In 2010, Qumo was released from Abu Salim as part of an amnesty program pressed by the U.S. Government as part of efforts to remove Libya from the state-sponsored terrorism list. In hindsight, opening the doors even once at Abu Salim was a disastrous policy mistake. A number of detainees were released in 2008 and 2010, and then in 2011, those former terrorists went back and broke out the rest of their terrorist colleagues at the start of the Libyan revolution.

In 2011, Qumo began establishing training camps on behalf of al-Qa'ida in Darnah, with many early trainees focused on supporting the al-Nusrah Front in Syria (also known as Jabhat Tahrir al-Sham, al-Qa'ida's main branch in the Levant). Also, as Abu Anas al-Libi set up the al-Qa'ida assassination units in Benghazi, Qumo was charged with the same role to establish the al-Qa'ida assassination units in Darnah. Qumo's right-hand leading the assassination units with him was Benghazi attacker Ali bin Taher. Ali was the terrorist who stole the CIA's armored sedan from the

U.S. Consulate during the attacks and then drove it home to Darnah. These units carried out near-daily assassinations from 2011 to mid-2014 when the LNA kicked off Operation Dignity. This was when the LNA declared war against terrorist groups in eastern Libya, and sought to wipe out al-Qa'ida's base of operations in Ganfouda, Benghazi.

On September 11th, 2012, Qumo was the leader of Ansar al-Sharia-Darnah (AAS-D). He sent approximately 15 to 20 fighters to attack the U.S. Consulate in Benghazi. The following AAS-D members were identified at the Consulate: Ali bin Taher, Hani al-Hawari, Anis al-Khurram, and Abdul Hamid al-Shaeri who was also affiliated with AQIM.

Separately, there was debate over the location of Qumo during the Benghazi attacks. We had long believed he was in Darnah based on local monitoring. However, unconfirmed reporting suggested he was in Benghazi during the attacks. We note this discrepancy for the record.

While the focus is usually on Qumo's Darnah-based activities, he also worked with terrorist Ahmed Ali Futaisi who controlled the city of Zliten, Libya. The two would facilitate terrorists from Tunisia through Sabratha, Libya, and then onward to the mountains near Darnah for training. This pipeline was funded using Libyan Government funds. Qumo and Futaisi received funds for each terrorist trained, and then were able to double-dip, as they then received funds for the same terrorist if he deployed to Syria. On just the back end of the transaction, having one trained terrorist to deploy to Syria will net you $5000. Funds were also provided by the Qatari and Turkish Governments. Two Libyan Intelligence Officers were also involved in this network. We are leaving their names out in case they are allied with the U.S. Government as our Government continues to choose the wrong side of the terrorism fight in Libya.

Shifting gears, starting in January 2014, Qumo had a major falling out with LIFG stemming from the incident when Ali Wanis Bukhamada, the son of Libyan Special Forces Commander Colonel Wanis Bukhamada, was kidnapped in Benghazi. The kidnapping was a joint operation between LIFG and Qumo's AAS-D, whereas AAS-D carried out the kidnapping. Even though LIFG was involved in the operation, they used the situation to their benefit and acted as if they assisted in releasing the son. The money that was supposed to go to Qumo to reimburse him for funding the operation was stolen by Benghazi attacker Ali bin Taher and his group at the time. This led to Ali's murder on April 7th, 2014, outside of Darnah. Qumo was incredibly upset over the murder of Ali, and it essentially ended his relationship with LIFG and al-Qa'ida going forward.

When ISIS started to make headway in Darnah in 2014, Qumo was the first member of AAS-D to pledge allegiance to them. Qumo was also essentially one of the key ISIS figures at the start in Darnah, like Benghazi attacker, Mahmoud al-Barassi was for ISIS in Benghazi around the same time. Qumo was all in with ISIS when they arrived in Darnah including personally providing the facilities initially used by the group. Even his sons fought with ISIS at the time. His youngest son, Ibrahim Sufyan Ibrahim bin Qumo was killed in Sirte, Libya, on August 12th, 2016, while fighting for ISIS. Another son, Faris Sufyan bin Qumo, just one month before his brother's death, was arrested enroute from Darnah to the Tobruk Airport and charged with terrorism.

In 2015, when the LNA started counterterrorism operations in Darnah against the terrorists, al-Qa'ida and ISIS tried to stay aligned against the LNA. However, the groups turned on each other and the terrorists had to choose which group they were more fundamentally aligned with. Qumo stayed with ISIS and was arrested for his support of the group in June 2015 by the DMSC, which was the al-Qa'ida arm in the city. He was able to get himself out of custody, possibly by pretending to pledge allegiance back to al-Qa'ida.

On June 9th, 2018, Egyptian al-Qa'ida leader, Omar Sorour, was killed during a series of air raids carried out by the LNA in Darnah. It had been reported that Qumo was either captured or killed in this incident. However, we have been unable to identify a post-mortem photograph of Qumo, nor have we been able to locate him in custody. If in LNA custody, due to his profile, we assume he would be in the Gernada Prison in al-Baydah, Libya. In contrast, a body of reporting suggests that Bin Qumo is at-large and, as of 2022, was located in Adana, Turkey. He reportedly had physically changed his appearance and was no longer recognizable via available historic photographs.

Benghazi: Know Thy Enemy

(18) Boka al-Oraibi, full name Mohammed Ibrahim Ahmed al-Arieby, from Benghazi. Boka had no jihadist past before the Libyan revolution. In 2011, Boka trained in Benghazi as a revolutionary and then fought in Ajdabiya before arriving on the front lines of the battles in Sirte. While fighting, he became affiliated with several Islamists and battle-hardened terrorists.

In September 2012, at the time of the U.S. Consulate and CIA Annex attacks, he was the Leader of Libya Shield Two. In this role, Boka assisted Wissam Bin Humaid (the Leader of Libya Shield One) and helped stall the CIA's rescue force coined "Team Tripoli" at the Benghazi International Airport. Libya Shield successfully kept Team Tripoli at the airport for over three and ½ hours. The delay at the airport assisted Wissam in setting up the deadly mortar strike that occurred at the Annex. Once Wissam was ready to co-locate all the Americans on the Annex, he gave Boka's Libya Shield Two militiamen permission to transport Team Tripoli from the airport to the Annex. Twelve minutes after Team Tripoli arrived at the Annex, it was struck by six precision mortar strikes.

Boka helped co-found the al-Qa'ida-affiliated BRSC with a number of the Benghazi attackers while leading the Brega Martyrs Brigade.

Starting in 2014, he participated in the Battle of Benghazi against the LNA. Later in 2014, Boka also pledged allegiance to ISIS and was the commander of one of its brigades based in Sirte called the Ibn al-Sheikh Brigade. In October 2014, "Guardians of Blood" burnt down Boka's house after he executed his cousin for serving in the LNA. Guardians was an organization of local families seeking revenge for the assassinations by terrorists of innocent civilians in Benghazi starting in 2011. On March 23rd, 2015, the LNA killed Boka during battles in the Bouatni district south of Benghazi.

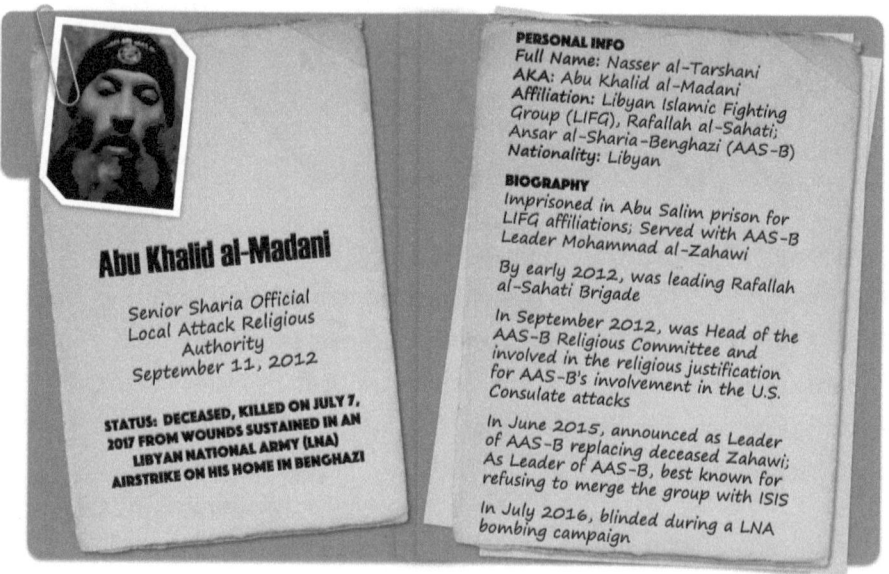

Abu Khalid al-Madani
Senior Sharia Official
Local Attack Religious Authority
September 11, 2012

STATUS: DECEASED, KILLED ON JULY 7, 2017 FROM WOUNDS SUSTAINED IN AN LIBYAN NATIONAL ARMY (LNA) AIRSTRIKE ON HIS HOME IN BENGHAZI

PERSONAL INFO
Full Name: Nasser al-Tarshani
AKA: Abu Khalid al-Madani
Affiliation: Libyan Islamic Fighting Group (LIFG), Rafallah al-Sahati; Ansar al-Sharia-Benghazi (AAS-B)
Nationality: Libyan

BIOGRAPHY
Imprisoned in Abu Salim prison for LIFG affiliations; Served with AAS-B Leader Mohammad al-Zahawi

By early 2012, was leading Rafallah al-Sahati Brigade

In September 2012, was Head of the AAS-B Religious Committee and involved in the religious justification for AAS-B's involvement in the U.S. Consulate attacks

In June 2015, announced as Leader of AAS-B replacing deceased Zahawi; As Leader of AAS-B, best known for refusing to merge the group with ISIS

In July 2016, blinded during a LNA bombing campaign

(19) **Abu Khalid al-Madani,** with real name Nasser al-Tarshani from Benghazi. Madani got his terrorism start in the LIFG. He was imprisoned in Abu Salim prison by the Gaddafi regime for being a member of LIFG. He was a close associate of detainee Mohammad al-Zahawi while in Abu Salim. He was released in 2011 and returned to Benghazi to fight in the Libyan revolution, and by early 2012, he was leading the Rafallah al-Sahati Brigade.

In mid-2012, he joined AAS-B and led the group's Sharia Committee by the end of 2012. He supported the religious justification for AAS-B's involvement in the U.S. Consulate in Benghazi attacks on September

11th, 2012. In 2013, when AAS-B started using the moniker Ansar al-Sharia-Libya (ASL), Zahawi became the ASL Leader for Libya, and Madani was sworn in as the ASL Leader for Benghazi.

In 2014, Madani was one of the BRSC founders with close associates Zahawi, Wissam bin Humaid and Mukhtar Abu Rizeza Ahad. Starting in 2014, Madani fought in the Battle of Benghazi against the LNA. On January 24th, 2015, Madani delivered an audio eulogy for Zahawi, who was killed in the Battle of the Benina Airport. Madani became Zahawi's replacement. As Leader of AAS-B, Madani was best known for refusing to merge the group with ISIS. His leadership likely led to the group's downfall, which disbanded in May 2017.

In July 2016, Madani was blinded during an LNA bombing campaign in the Chinese Building neighborhood in downtown Benghazi. He then fled to the al-Sabri neighborhood. On July 7th, 2017, Madani was killed in his home when the LNA bombed it.

Chapter 7

Enemies: Brothers in Arms—The Attackers

"Brothers in Arms" is the listing of attackers on the ground over the course of the six attacks that occurred against us on September 11th and September 12th, 2012. Attackers are listed alphabetically by family name. During our investigation, the following terrorists were confirmed to be directly involved and present at the U.S. Consulate attacks in Benghazi. Those suspected of being at the CIA Annex attacks on September 12th are labeled as such. To remain focused on the attackers, please note that no attack planners, communications conduits, facilitators, financiers, or trainers are listed in the "Brothers in Arms" section.

The Benghazi attacks were a family affair—it was not just family members by blood which there are many in the following pages, but brotherhoods formed in wars, be it in Sudan, Afghanistan, Algeria, Iraq, Libya, Mali, and Syria. The attackers had decades of bonds with one another—they were a family which yes, became broken as they were pitted against each other in the years following the arrival of ISIS in Libya. Al-Qa'ida battled ISIS for primacy in the country, luckily they both lost. Still, they were all brothers nonetheless on the night of September 11th, and they got away with taking our brothers from us.

A

- (20) Hamdi Abbasi, full name Hamdi Ali Abu Bakr Abbasi
- (21) Adel Ahmad al-Abdali
- (22) Rashid Abdullah
- (23) Hafiz al-Aqouri, full name Hafiz Salem Zaid al-Aqouri
- (24) Khaled al-Ammari, full name Khaled Masoud Saeed al-Ammari
- (25) Mohammad al-Ammari, full name Mohammad Masoud Saeed al-Ammari
- (26) Muftah al-Ammari, full name Muftah Masoud Saeed al-Ammari

(27) Tariq Amitiq, full name Tariq Mohammad Salem Amitiq
(28) Marei al-Arfi, full name Marei Saleh Mohammad Hamad al-Arfi
(29) Imad al-Awami, full name Imad Mohammad Abdullah al-Awami
(30) Mahmoud al-Awami, full name Mahmoud Idris Musa al-Awami
(31) Taher al-Awami, full name Taher Mustafa Mohammad al-Awami
(32) Yousef al-Awami, full name Yousef Idris Musa al-Awami
(33) Mohammed al-Ayouni, full name Mohammed bin Salem al-Ayouni
(34) Abdel-Qader Azzouz, full name Abdel-Qader Abdel-Salam Abdel-Qader Azzouz
(35) Jaafar Azzouz, full name Jaafar Abdel-Salam Abdel-Qader Azzouz

B

(36) Zubayr al-Bakoush, full name Zubayr Hassan Omar al-Bakoush
(37) Ziad Balaam, full name Farid Mohammad Muhammed Balaam
(38) Mahmoud al-Barassi, full name Mahmoud Ali al-Barassi
(39) Marwan al-Barki, full name Marwan Hassan al-Barki
(40) Omar al-Barki, full name Omar Mohamed Zayed al-Barki
(41) Ayman Bouamoud, full name Ayman Faraj Hassan Bouamoud al-Bargathi
(42) Walid al-Barnawi, full name Walid Abdel Qader Mohammed al-Barnawi
(43) Salim Bayou, full name Salim Mustafa Nasser Bayou
(44) Ahmed Abu Khatallah, full name Ahmed Faraj Salem Boukhattala
(45) Abdullah Bouzkia, full name Abdullah Mohamed Mohamed Bouzkia
(46) Ahmed Buhajar, full name Ahmed Mukhtar Buhajar

D

(47) Mohamed Ben Dardaf, full name Mohammad Mohammad Mahmoud bin Dardaf
(48) Youssef al-Darsi, full name Youssef Ibrahim Mohammed al-Darsi

F

(49) Hamad al-Fakhri, full name Hamad Noah Younis al-Fakhri
(50) Fawzi al-Faydi, full name Fawzi Mahmoud Ahleel al-Faydi
(51) Khaled al-Faydi, full name Khaled Ali Salem al-Faydi

G

(52) Imad al-Gharyani, full name Imad Mohammad Mohammad al-Gharyani

(53) Mahdi Saad al-Ghaythi, full name Al-Mahdi Saad al-Mahdi Boulbeid al-Ghaythi

H

(54) Abu Hamza al-Tabawi, real name Ali Mohammad al-Toghi Hammadi

(55) Suhaib al-Hamroush, full name Suhaib Awad al-Hamroush

(56) Faisal bin Hariz, full name Faisal Mohammad Ali bin Hariz

(57) Hudhayfah bin Hariz, full name Hudhayfah Mohammad Ali bin Hariz

(58) Talal bin Hariz, full name Talal Mohammad Ali bin Hariz

(59) Ali Ouni al-Harzi, full name Ali bin al-Tahar bin al-Falih al-'Awni al-Harzi

(60) Faraj Shakku, real name Faraj Musa al-Hasnawi

(61) Ahmed Hazaa, full name Ahmed el-Sayed Hazaa Hassan

(62) Hani al-Hawari

(63) Anis al-Houti, full name Anis Faraj Abd al-Salam Mohammad al-Amin al-Houti

I

(64) Mustafa al-Imam, full name Mustafa Mohamed al-Imam

J

(65) Mohammad Jaafar, full name Mohammad Hassan Rajab Jaafar

(66) Abdel Moneim al-Jahawi, full name Abdel Moneim Ibrahim al-Jahawi variant el-Gahhawy

K

(67) Munther al-Kubti, full name Munther Mansour al-Kubti

(68) Atef al-Karami, full name Atef Mustafa Mohammad al-Karami

(69) Hassan al-Karami, full name Hassan Mohammad Abdullah al-Karami

(70) Ahmed bin Nasser Karim

(71) Ali al-Karshini, full name Ali Mohammad Mohammad Ibrahim al-Karshini

(72) Xris Katsiroubus, full name Xristos Nikolaos Katsiroubus

(73) Abu Yaqin al-Libi, real name Mohammad Fawzi Al-Senussi al-Kawafi

(74) Mohammad al-Kawil, full name Mohammad Hamad Ali al-Kawil

(75) Amin Kelfa, full name Amin Ali Meloud Kelfa

(76) Marei Zoghbi, full name Marei Abdel-Fattah Khalil

(77) Malik al-Khazmi, full name Al-Sadiq Salem al-Khazmi

(78) Anis al-Khurram, full name Anis Abdel Salam al-Khurram

(79) Alaa al-Kilani, full name Alaa el-Din Salem Hassan al-Kilani

L

(80) Salem al-Lawati, full name Salem Yassin Ali al-Lawati

M

(81) Ahmed al-Majbari, full name Ahmed Abdel Rahim al-Majbari

(82) Muharib al-Majbari full name Muharib Musa Mohammad al-Majbari

(83) Jalal Makhzoum, full name Jalal Ahmed Mohammad Makhzoum

(84) Marei al-Manfi, full name Marei Mohammed Hassan al-Manfi

(85) Mohammad al-Manfi, full name Mohammad Abd al-Salam al-Manfi

(86) Mohammad Jaber Abd al-Maqsoud, full name Mohammad Jaber Abd al-Wahhab Abd al-Maqsoud

(87) Hamza al-Darnawi, real name Abd al-Hamid Mohammad Abd al-Hamid Masli

(88) Ali Medlej

(89) Asmi Ahmed, full name Asmi Ahmed Mohamedin

(90) Ahmed al-Munfi, full name Ahmed Murtaha Mukhtar al-Munfi

(91) Ahmed al-Mushaiti, full name Ahmed Hassan al-Sharif Mohammed al-Mushaiti

(92) Majdi al-Mushaiti, full name Majdi al-Maliq al-Ghanai al-Mushaiti

(93) Salah al-Mushaiti, full name Salah al-Din Ali Ibsekri al-Mushaiti

(94) Shoaib al-Mushaiti, full name Shoaib Ali Ibsekri Shoaib al-Mushaiti

O

(95) Abdulaziz al-Obaidi, full name Mohammad Abdulaziz al-Mahdi al-Obaidi

(96) Abu Zaid al-Shalawi, full name Abu Zaid Mohammad al-Shalawi al-Obaidi

(97) Osama al-Obaidi, full name Osama Abdullah al-Ghaythi al-Obaidi

R

(98) Karim Moawad al-Rahmani, full name Karim Moawad Mohammad al-Rahmani

(99) Ramadan al-Rubaie, full name Ramadan Mohammad al-Rubaie

S

(100) Yunus Emre Sakarya

(101) Hamza al-Sallak, full name Muftah Hamza al-Sallak

(102) Khaled al-Saqzli, full name Khaled Abdelwahab Mustafa al-Saqzli

(103) Mansour al-Shaalali, full name Mansour Juma al-Shaalali

(104) Abdul Hamid al-Shaeri, full name Abdul Hamid Saad Abdul Karim al-Shaeri

(105) Hassan al-Shukri, full name Hassan Musa al-Kilani al-Shukri

(106) Abdul-Ati Abu Sitta, full name Abdul-Ati al-Shtiwi Abu Sitta

(107) Emad al-Shuqabi, full name Emad Faraj Mansour al-Shuqabi

(108) Majdi al-Suwaie, full name Majdi Abdel Moneim Mohammad al-Suwaie

(109) Ali bin Taher, full name Ali Abdullah Muftah bin Taher

(110) Ahmed al-Tajouri, full name Ahmed Ibrahim Attia Hamad al-Tajouri

(111) Islam al-Tarhouni, full name Islam Ahmed al-Muntasir Belaid al-Mahdawi al-Tarhouni

(112) Ramadan Trabelsi, full name Ramadan Tawfiq Abdullah Trabelsi

U

(113) Faraj al-Chalabi, full name Faraj Husayn Hasan al-Shalabi al-Urfi

(114) Mansour al-Urfi, full name Mansour Abdel Salam al-Tariki al-Urfi

W

(115) Mahmoud al-Wahishi, full name Mahmoud Saad Hussein Abdul Ghani al-Wahishi

Enemies: Brothers in Arms—The Attackers

(116) Mohammad al-Werfalli, full name Mohammad Ali Masoud bin Jaber al-Werfalli

Z

(117) Idris al-Zawi, full name Idris Othman Abu Bakr al-Zawi

(118) Jamaica, real name Yahya Abdel-Syed Saleh al-Zawi

(119) Hussam bin Hassouna al-Zaytouni, full name Hussam bin Hassouna Abdullah al-Zaytouni

A

(20) Hamdi Abbasi, full name Hamdi Ali Abu Bakr Abbasi from Benghazi. Abbasi was a member of the Martyrs Group. The Martyrs Group, also referred to in the past as the Islamic Martyrs Movement, was started in Libya after the 1996 Abu Salim prison massacre and the resulting uprisings. The group's first high-profile terrorist act was the attack on the Egyptian Consulate in Benghazi on April 5th, 1996, killing one guard. They also led one of the two failed assassination attempts against Gaddafi in the late 1990s.

The leaders who established the Martyrs Group were Mohammad

al-Hami, Hamza Bouchertila, and Mohammed Saleh Mohammed al-Makawi. As of 2022, the Martyrs Group was still in existence, and it was a covert operations group with Benghazi attacker Mohammad al-Kawil as one of its senior leaders. In addition to Abbasi and Kawil, the following additional Benghazi attackers were at one-point members of the Martyrs Group: Ayman Bouamoud, Muftah al-Ammari, Mohammad al-Ammari, Mohammed al-Gharabi, Atef al-Karami, and Salah al-Mushaiti. The Martyrs Group reportedly also had a branch in Europe.

Abbasi had been a prisoner in Abu Salim prison. In 2011, he traveled to fight in Misrata against Gaddafi forces with former Abu Salim prisoners and future Benghazi Consulate attackers Mohammad al-Zahawi, Mohammad al-Manfi, and Ahmed al-Mushaiti; as well as with terrorist Mohammad al-Ferjani.

After fighting in Misrata, Abbasi returned to Benghazi, joining close associate Zahawi's new AAS-B group, and was present at the group's first annual Sharia conference in June 2012 in Benghazi. In September 2012, he participated on behalf of AAS-B in the U.S. Consulate attacks in Benghazi. Abbasi took on the role of being a coordinator for wounded terrorists and traveled to assist them not just in eastern Libya but in both Syria and Turkey. As part of assisting in the care of injured terrorists, he also facilitated their travel and provided false documents for travel. In one example, after the battle in Sirte, against ISIS, Tunisian terrorist and prisoner Khaled Ali al-Zoubi reported that Abbasi provided him a Libyan passport under the name Mustafa Ali Mustafa to travel to Libya to fight with.

In June 2015, Abbasi was arrested in Misrata. A number of official letters from the Libyan Government poured in requesting his release. He remained in custody and was handed over to the RADA Special Deterrence Force in Tripoli. As of 2022, if still detained, he would likely be located in the Mitiga prison in Tripoli.

(21) Adel Ahmad al-Abdali, alias Abu Zubair al-Libi, from Benghazi. Starting in the early 2000s, Adel was a member of AQI, in what was referred to as the "al-Zarqawi cell"—a nod to AQI's Leader Abu Musab al-Zarqawi. His primary role was facilitating Libyan fighters to Iraq to fight against Americans. During one of his border crossings between Syria and Libya, he was detained and extradited to Libya. Adel served in Abu Salim prison from 2004 until 2011 when terrorists freed him during the Libyan revolution.

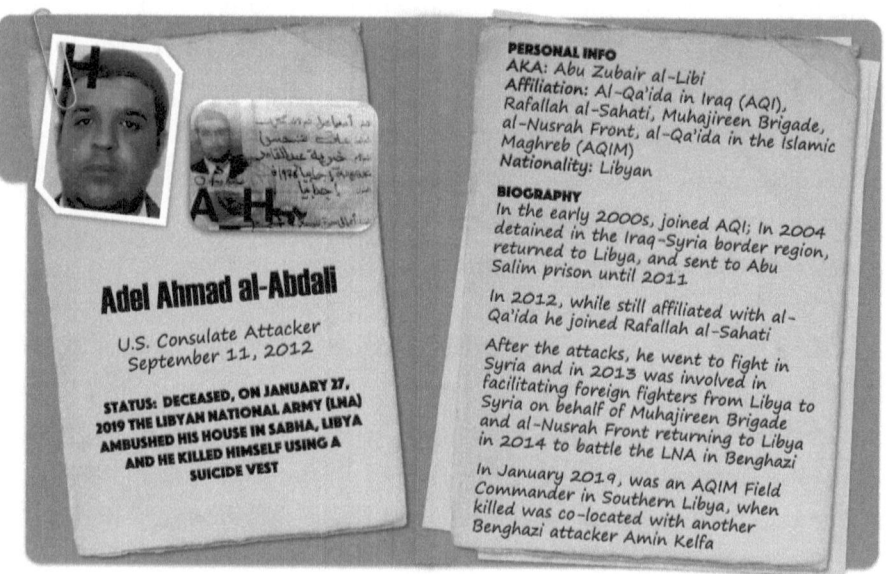

Adel Ahmad al-Abdali
U.S. Consulate Attacker
September 11, 2012

STATUS: DECEASED, ON JANUARY 27, 2019 THE LIBYAN NATIONAL ARMY (LNA) AMBUSHED HIS HOUSE IN SABHA, LIBYA AND HE KILLED HIMSELF USING A SUICIDE VEST

PERSONAL INFO
AKA: Abu Zubair al-Libi
Affiliation: Al-Qa'ida in Iraq (AQI), Rafallah al-Sahati, Muhajireen Brigade, al-Nusrah Front, al-Qa'ida in the Islamic Maghreb (AQIM)
Nationality: Libyan

BIOGRAPHY
In the early 2000s, joined AQI; In 2004 detained in the Iraq-Syria border region, returned to Libya, and sent to Abu Salim prison until 2011

In 2012, while still affiliated with al-Qa'ida he joined Rafallah al-Sahati

After the attacks, he went to fight in Syria and in 2013 was involved in facilitating foreign fighters from Libya to Syria on behalf of Muhajireen Brigade and al-Nusrah Front returning to Libya in 2014 to battle the LNA in Benghazi

In January 2019, was an AQIM Field Commander in Southern Libya, when killed was co-located with another Benghazi attacker Amin Kelfa

In September 2012, Adel was a member of al-Qa'ida and the Rafallah al-Sahati Brigade (which he joined during the revolution) when he took part in the attacks on the U.S. Consulate in Benghazi. After the attacks, he fled to fight in Syria. He joined Jaish al-Muhajireen wal-Ansar (also known as the Muhajireen Brigade), a group led by Abu Talha al-Libi with real name Abdul Moneim al-Hasnawi. Hasnawi went on to be a Senior Leader in AQIM. Adel also joined local al-Nusrah Front efforts in Syria and was a member when it transitioned into al-Qa'ida in Syria starting in approximately 2013. Like his prior support to AQI during the U.S. invasion of Iraq, Adel facilitated fighters from Libya to Syria.

After the LNA began counterterrorism operations in Benghazi in 2014, Adel returned to Libya to fight with al-Qa'ida. When terrorists lost most of Benghazi in 2017, he fled to Misrata, Libya. He then fled Misrata for Sabha, Libya, where he lived under a fake name, Ismail Mohammad

Ali Hassan, and served as a Field Commander for AQIM. He was tracked down and located by the LNA. On January 27th, 2019, LNA ambushed his house in Sabha and a firefight ensued. Adel was wearing an explosive vest and blew himself up when LNA closed in on him. He killed his mother and daughter during the suicide bombing, and his wife survived with extensive injuries. Of note, another Benghazi attacker, Amin Kelfa, was captured in the raid that killed Adel.

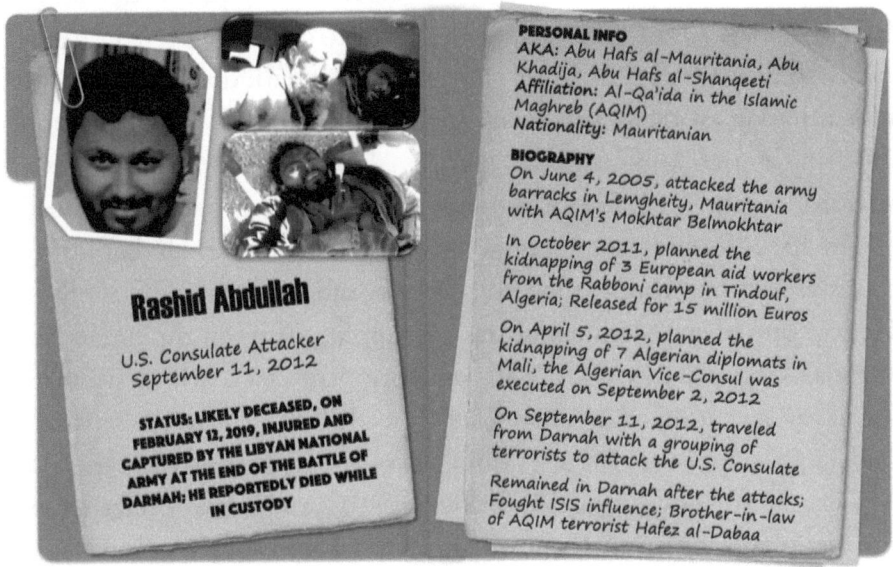

(22) **Rashid Abdullah,** with aliases Abu Khadija, Abu Hafs al-Mauritanian, and Abu Hafs al-Shanqeeti, from Mauritania. Rashid was a member of AQIM, where he was one of the legislators of the terrorist group. He was a close associate of Benghazi attack Mastermind MBM. He participated in the June 4th, 2005, attack at the army barracks in Lemgheity, Mauritania, which was planned by MBM. Fifteen Mauritanian soldiers and nine attackers were killed.

In 2011, Rashid traveled to Libya from Algeria to support the Libyan revolution. He was well-connected to Darnah-based terrorists and was a close associate of Sufyan bin Qumo, the Leader of AAS-D, and Egyptian terrorist Hesham Ashmawy. Rashid was also married to the daughter of Youssef Abdel Karim al-Hasadi and the sister of fellow AQIM member

Hafez al-Dabaa with alias Abu Ayoub.

On May 17th, 2011, Dabaa was one of two Libyan AQIM members arrested by Tunisian counterterrorism officials for carrying out the first terrorist attack in Tunisia following the fall of dictator Zine el-Abidine Ben Ali. Dabaa was serving a 20-year sentence in a Tunisian prison when two Tunisian diplomats were kidnapped in Tripoli, Libya, on March 21st, 2014, and April 17th, 2014. Tunisian President Moncef Marzouki was forced to release Dabaa and terrorist Emad al-Lauwaj Badr with alias Abu Jafar al-Libi as a hostage exchange in April 2014.

Rashid was also involved in planning the kidnapping of three humanitarian workers, two from Spain and one from Italy, in October 2011 in the Rabboni camp in Tindouf, Algeria. The three hostages were released in July 2012, for a ransom of approximately 15 million Euros.

In 2012, Rashid issued several terrorist fatwas in Darnah that justified the assassinations against the army, security officials, and judicial officers. He was one of the planners of the kidnapping of seven diplomats from Algeria's Consulate in Gao, Mali, on April 5th, 2012, which ultimately led to the execution of Algerian Vice-Consul, Taher Touati on September 2nd, 2012. In June 2012, Rashid survived an ambush by the Arab Movement of Azawad that was formed in 2012 and initially referred to as the National Liberation Front of Azawad, organized by secular Arabs to battle the Taureg tribe in Timbuktu, Mali.

On September 11th, 2012, he traveled with a group of terrorists from Darnah to participate in the attacks on the U.S. Consulate in Benghazi. In the years following the attacks, Rashid split from MBM's Battalion and joined the Abu al-Waleed Desert group led by Abu al-Walid al-Sahrawi. Al-Sahrawi defected from MBM's Battalion as he wanted to be more aligned with ISIS, which MBM would never do due to his devotion to then al-Qa'ida leader Dr. Ayman al-Zawahiri.

In the 2014 to 2015 timeframe, Rashid was heavily involved in the clashes between al-Qa'ida and ISIS for control of Darnah. He would respond on behalf of al-Qa'ida's affiliates to fatwas issued by Abu al-Bara al-Azdi from ISIS. Abu al-Bara, through ISIS sharia lawmakers, was trying to justify that those groups affiliated with al-Qa'ida were infidels.

Rashid also received direct funding from the Government of Qatar, according to terrorist associate Hesham Ashmawy in debriefings after his detention. The funding was to support efforts with the Government of Turkey, presumably for the foreign fighter pipeline into Syria.

From May 7th, 2018, through February 12th, 2019, the LNA battled in the military campaign coined the Battle of Darnah to rid the city of terrorists, primarily al-Qa'ida's DMSC. On the last day of the war, on February 12th, Rashid was injured when he was captured in LNA sweeps netting 80 terrorists at the end of the battle. Rashid was likely deceased. Rashid's brother-in-law Dabaa, who was previously the Spokesman for the DMSC, was also taken into custody.

(23) Hafiz al-Aqouri with full name Hafiz Salem Zaid al-Aqouri with alias Hayaka Alla from Ben Younis, Benghazi. He had a historical relationship with al-Qa'ida operative Abdul Basit Azzouz. In the 1990s, he traveled to Afghanistan to fight with al-Qa'ida. In Afghanistan, he established a close relationship with Dr. Ayman al-Zawahiri. In 2011, during the revolution, he fought with Wissam bin Humaid's militia, Free Libya Martyrs Brigade but stayed a member of al-Qa'ida. In spring 2011, while fighting in the Battle of Brega, Aqouri was injured and had two of his fingers amputated.

That same year, Zawahiri asked Aqouri to help facilitate 400 al-Qa'ida fighters to Libya for the revolution through the southern border with Sudan. The al-Qa'ida fighters were smuggled to Libya to help overthrow the Gaddafi regime. Aqouri was not the only al-Qa'ida member tasked with facilitating al-Qa'ida members into Libya during the revolution. Other al-Qa'ida members tapped by Zawahiri included Abdel Moneim al-Madhouni; and Salem Nour al-Din al-Dibsky variant Nur al-Din al-Dibiski, with aliases Abu al-Ward and Attiyah Jalil, who took control of the city of al-Khums, Libya.

In 2012, Aqouri joined Libya Shield One under Wissam bin Humaid. The Libya Shield Forces were funded by the Government of Libya. From February to July 2012, Aqouri, Wissam, and Yahya al-Maqsabi were directly involved in the Kufra Conflict in southeast Libya between the Toubou and Zawiya tribes. Aqouri was a Zawiya tribesman. Over 100 people were killed in the conflict in which Libya Shield One engaged in battles.

Aqouri, again a long-time Zawahiri associate, was assessed to have played a role in the September 11th, 2012, attack on the U.S. Consulate in Benghazi. In 2012, during the attacks, Aqouri was doing double duty within Libya Shield One, acting as the group's Spokesman and serving as a Deputy to Leader Wissam. He shared the Deputy role with Wissam's brother Qais who handled the business side, while Aqouri led the operational side. Wissam's other brother, Mohammad, was also involved in the day-to-day activities of Libya Shield One, but he was a fighter. We confirmed that at least three members of Libya Shield One participated directly in the attack on the Consulate, including attackers Mohamed Ben Dardaf, Emad al-Shuqabi, and Ahmed Buhajar.

It was the attack on the CIA Annex in Benghazi on September 12th, 2012, that we assess Libya Shield One played a direct role and successfully pulled off their hidden hand in the attacks' involvement. As such, Aqouri was a prime suspect in orchestrating the deadly mortar attack on the Annex that killed CIA GRS Officers Rone and Bub. As of 2022, Aqouri was at-large.

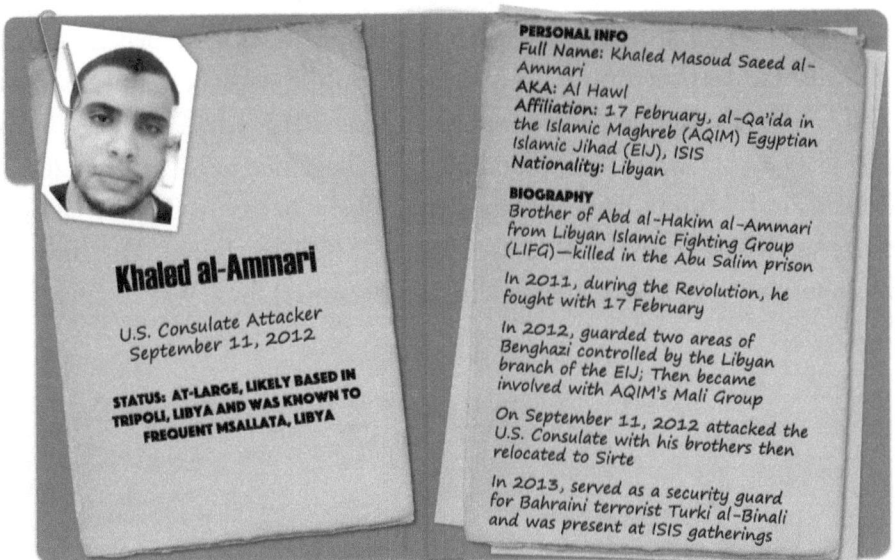

(24) **Khaled al-Ammari,** full name Khaled Masoud Saeed al-Ammari with alias Al Hawl from Benghazi. Note this was a separate individual from terrorist Khaled Abu al-Qasim Ahmed al-Ammari. In 2011, during the Libyan revolution, he fought with 17 February.

Khaled was from a family of jihadists. His older brother was Abd al-Hakim al-Ammari with alias Abu Muslim. He was a senior member of the LIFG's Security Committee and was killed during the Abu Salim prison massacre in 1996. Two brothers, Muftah al-Ammari and Mohammad al-Ammari were members of the Martyrs Group in Libya, which was a covert group similar to the al-Qa'ida's Khorasan group. Another brother was Abdul-Ghani al-Ammari, who first fled to Qatar, where he lived and traveled under the name Ahmed Nasser al-Shaheen. Abdul-Ghani worked out of the Naim bin Rabia Mosque in Qatar before relocating to the United Kingdom.

In 2012, Khaled was guarding several areas in Benghazi controlled by the Libyan faction of the EIJ in the 602 area and the Ard Alherasaa neighborhood, where Benghazi's primary Egyptian market was located. He was also involved in several assassinations in these two locations. On September 11th, 2012, he attacked the U.S. Consulate in Benghazi with his terrorist brothers Mohammad and Muftah al-Ammari.

After participating in the September 2012 attacks, Khaled relocated to Sirte. He was seen quite frequently in public with Bahraini terrorist Turki al-Binali after he arrived in Sirte in June 2013. Khaled was acting as a security escort for Turki. Khaled pledged allegiance to ISIS. As of 2022, Khaled and Muftah were at-large, and likely based in Tripoli, Libya, however, Khaled was known to frequent Msallata, Libya. His brother Mohammad was killed fighting in the Second Libya War against the LNA.

Mohammad al-Ammari

U.S. Consulate Attacker
September 11, 2012

STATUS: DECEASED, KILLED IN THE BATTLE OF BENGHAZI BY THE LIBYAN NATIONAL ARMY (LNA)

PERSONAL INFO
Full Name: Mohammad Masoud Saeed al-Ammari
Affiliation: Rafallah al-Sahati Brigade, Martyrs Group
Nationality: Libyan

BIOGRAPHY
From a jihadist family; Former Abu Salim prisoner along with two brothers Abd al-Hakim and Muftah; Freed prior to the Libyan Revolution

In 2011, during the Revolution, joined Rafallah al-Sahati; Involved in assassinating police and security officials in the 602 Neighborhood in Benghazi

On September 11, 2012 he and two brothers Khaled and Muftah carried out the U.S. Consulate in Benghazi attacks; Both Khaled and Muftah are at-large in western Libya

In 2014, fought with terrorists against the LNA in the kick-off of the Second Libyan War and was killed

(25) Mohammad al-Ammari, full name Mohammad Masoud Saeed al-Ammari from Benghazi, was a member of the Martyrs Group. He became involved in terrorism after being influenced by the extremist ideology of his older brothers, Qatari/British resident Abdel Hakim al-Ammari and terrorist Abdel Ghani al-Ammari. However, he became more extreme than both of them. He was detained in the past for involvement in terrorism, and he served in the Abu Salim prison, where his brother Abdul Ghani was killed in 1996. He was released before the Libyan revolution.

In 2011, during the revolution, he joined the Rafallah al-Sahati Brigade. His primary activity for Rafallah al-Sahati was providing security for the residential area within the 602 Neighborhood in Benghazi; in reality, he was assassinating civilians who had chosen to join different police and security services in the city. On September 11th, 2012, Mohammad attacked the U.S. Consulate with two additional brothers, Khaled, who was 17 February at the time, and Muftah, who was also part of the Martyrs Group. Muftah had also joined the Rafallah al-Sahati Brigade with Mohammad but then, in 2012, moved on to AAS-B. Mohammad was killed by the LNA during the Battle of Benghazi.

Muftah al-Ammari

U.S. Consulate Attacker
September 11, 2012

STATUS: AT-LARGE, LIKELY BASED IN WESTERN LIBYA AND HAS STRONG TRIBAL AND TERRORIST TIES TO AL-AMAMRA, LIBYA

PERSONAL INFO
Full Name: Muftah Masoud Saeed al-Ammari
Affiliation: Martyrs Group, Rafallah al-Sahati Brigade, Ansar al-Sharia-Benghazi (AAS-B)
Nationality: Libyan

BIOGRAPHY
Prior to the 2011 Libyan Revolution, detained for being a Martyrs Group terrorist in Abu Salim prison where his brother Abd al-Hakim al-Ammari died in the 1996 massacre

In 2011, during the Revolution, fought with Rafallah al-Sahati and in 2012 joined AAS-B where he led assassinations against internal security officers in Benghazi's 602 District

On September 11, 2012 attacked the U.S. Consulate in Benghazi with his two brothers

Starting in 2014, fought the Libyan National Army (LNA) during the Battle of Benghazi and by 2017 had fled to western Libya

(26) Muftah al-Ammari, full name Muftah Masoud Saeed al-Ammari from Benghazi. Muftah al-Ammari (and his grouping of terrorist brothers) were mentioned previously in the biographies of his brothers, Benghazi attackers Khaled and Mohammad al-Ammari. Muftah was previously detained at the Abu Salim Prison on terrorism charges for being a member of the Martyrs Group. In 2011, during the Libyan revolution, he fought with the Rafallah al-Sahati Brigade. Then in 2012, he joined AAS-B. He operated as a terrorist for AAS-B in the 602 District of Benghazi, where his family of terrorists resided. His primary focus was targeting and assassinating internal security officers. On September 11th, 2012, he attacked the U.S. Consulate in Benghazi with Khaled and Mohammad.

Muftah fought against the LNA during the Battle of Benghazi from 2014 to 2017. He then fled to western Libya and went off the grid. A likely hiding location for him was near al-Amamra, Libya, where his tribe originated. This area had a concentration of families who joined terrorist groups and fought in the city of Benghazi.

(27) Tariq Amitiq, full name Tariq Mohammad Salem Amitiq from Benghazi. Tariq was not known to have any involvement in terrorism before the Libyan revolution in 2011. In 2012, he joined AAS-B and was a member of the Al-Ansar Mosque cell in the al-Hadayek neighborhood in Benghazi. As a day-to-day activity from 2012 to 2013, he ran a traveling roadshow across different neighborhoods in Benghazi, setting up tents and advocating for public support for AAS-B. Tariq was also a follower of Sheikh Osama al-Sallabi, who was a leader in the Muslim Brotherhood movement and the brother of Ismail al-Sallabi.

On the day of the U.S. Consulate attacks on September 11th, 2012, Tariq was one of the attackers that staged at attacker Mahmoud al-Awami's house in the Bouzgheba area of Benghazi. The attackers then traveled together from the house to stage near the U.S. Consulate for the evening attack. In 2014, as the LNA kicked off its counterterrorism operations in Benghazi, Tariq fled to Misrata, Libya. In Misrata, he joined the al-Samoud Brigade led by Salah Badi. As of 2022, Tariq was at-large and still operating in the city of Misrata.

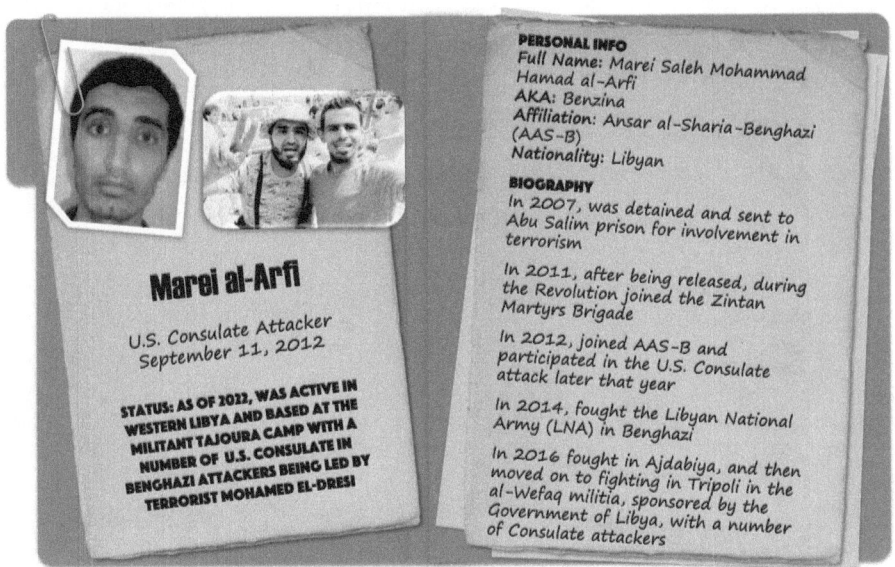

Marei al-Arfi
U.S. Consulate Attacker
September 11, 2012

STATUS: AS OF 2022, WAS ACTIVE IN WESTERN LIBYA AND BASED AT THE MILITANT TAJOURA CAMP WITH A NUMBER OF U.S. CONSULATE IN BENGHAZI ATTACKERS BEING LED BY TERRORIST MOHAMED EL-DRESI

PERSONAL INFO
Full Name: Marei Saleh Mohammad Hamad al-Arfi
AKA: Benzina
Affiliation: Ansar al-Sharia-Benghazi (AAS-B)
Nationality: Libyan

BIOGRAPHY
In 2007, was detained and sent to Abu Salim prison for involvement in terrorism

In 2011, after being released, during the Revolution joined the Zintan Martyrs Brigade

In 2012, joined AAS-B and participated in the U.S. Consulate attack later that year

In 2014, fought the Libyan National Army (LNA) in Benghazi

In 2016 fought in Ajdabiya, and then moved on to fighting in Tripoli in the al-Wefaq militia, sponsored by the Government of Libya, with a number of Consulate attackers

(28) Marei al-Arfi, full name Marei Saleh Mohammad Hamad al-Arfi with alias Benzina from Benghazi. Before the Libyan revolution, Marei served in Abu Salim prison after being detained in 2007. In 2011, during the revolution, he joined the Zintan Martyrs Brigade. In 2012, he joined AAS-B and participated in the U.S. Consulate attack later that year. He went on to fight with AAS-B in the al-Laythi area of Benghazi against the LNA. Marei was a wanted terrorist in several assassination cases in Benghazi.

In 2016, after al-Qa'ida affiliates were forced out of Benghazi by LNA's counterterrorism operations, Marei and fellow Benghazi attackers Faraj Shakku and Ahmed al-Tajouri led a sizable attack on the city of Ajdabiya, Libya. This attack was on behalf of the Benghazi Defense Brigades (BDB). Marei then relocated to fight in western Libya. In Tripoli, he joined the al-Wefaq militias with some of the Benghazi attackers in support of the Government of Libya against the LNA. On June 5th, 2020, Abdul Malik al-Madani, a spokesman for the forces of the Tripoli-based Government

Enemies: Brothers in Arms—The Attackers

of National Accord (GNA), posted pictures on Twitter with Benghazi attacker Marei as they celebrated an LNA defeat in Tripoli.

As of 2022, Marei was still active in western Libya and was based at the Tajoura camp, located southeast of Tripoli. The press would like you to believe this camp is a refugee center, vice the truth that it is a military-style camp hosting terrorists and funded by the Governments of Libya and Turkey. Marei was co-located there with several U.S. Consulate in Benghazi attackers. These Benghazi terrorists were being led at Tajoura by terrorist Mohamed el-Dresi with alias Mohammad al-Nass.

(29) Imad al-Awami, full name Imad Mohammad Abdullah al-Awami with aliases Abu Dujana and al-Maarsh, from Benghazi. Before the 2011 Libyan revolution, Imad was not reported to have been involved in terrorism. In 2011, during the Libyan revolution, he joined 17 February.

In the aftermath of the revolution, Imad set up an operational cell within 17 February that targeted military officers and security officials in Benghazi for assassinations. The terrorists in his cell included: Abdul Qader al-Misrati, full name Abdul Qader al-Misrati Ali al-Awjali with aliases Aqdoura and Abu Khaitham al-Libi, and Benghazi attackers Abdulaziz al-Obaidi and Khaled al-Ammari. Abdul Qader had been associated with a number of the Benghazi attackers. Abdul Qader had joined ISIS and the Ummah Brigade in Syria, where he carried out a suicide vehicle-borne improvised explosive device (SVBIED) attack on the same date as the CIA Annex attacks on September 12th, 2012. He killed 150 Syrian Army members in Saraqib City.

After Imad carried out the September 11th, 2012, attacks on the U.S. Consulate in Benghazi, no further information was available on him.

(30) Mahmoud al-Awami, full name Mahmoud Idris Musa al-Awami from Benghazi. Mahmoud was a member of AAS-B and had a close relationship with the Rafallah al-Sahati Brigade in early 2012. He was affiliated with the al-Ansar Mosque (and the al-Ansar Mosque Cell), and the Shuail Mosque. In 2012, he operated a traveling road show in Benghazi, where he set up tents in different neighborhoods and used his persuasion skills to convince young recruits to join AAS-B. Mahmoud was somewhat of a renaissance man as while he recruited fighters for jihad in Syria; he was also a journalist and poet.

Throughout 2012, Mahmoud assisted the facilitation of terrorists into the city, including Egyptian terrorists. Of note, his mother was Egyptian with her roots in the city of Damanhour, Egypt. As such, Mahmoud had long maintained contacts with extremists in Egypt. As additional background, from 2012 to 2013, the Awami's house was used as a reception center when foreign Islamist religious clerics visited Benghazi. He reportedly housed clerics from Yemen, including those affiliated with AQAP.

On September 10th, 2012, Mahmoud allowed attackers who traveled into Benghazi for the attacks to bed down at his home. This included Egyptian attackers funded by AQAP. Then on September 11th, Mahmoud's house was one of the launch points for terrorists to travel from enroute to the U.S. Consulate attacks. He and his brother Yousef, a fellow Benghazi attacker, started guarding the two street entrances in front of their family home in the Buzgheba area near the Shuail Mosque beginning in the early hours of September 11th up until the attackers departed in the evening hours. Throughout the day, attackers were granted access to the area through makeshift checkpoints outside the family home. Armed attackers were then visible entering the property, where they remained until they loaded up in vehicles after sundown on the night of the attacks.

He fled from Benghazi to western Libya with his family after the LNA started counterterrorism operations in Benghazi in 2014. The family settled in Qara Boli, Libya. As of 2022, Mahmoud was still at-large and was active in western Libya. On his social media, he posted himself greeting close associate Al-Sadiq al-Gharyani, the Grand Mufti of Libya, after a return medical trip from Turkey. Mahmoud was also close to Osama al-Sallabi, the brother of Ismail al-Sallabi.

(31) Taher al-Awami, full name Taher Mustafa Mohammad al-Awami from Benghazi. Taher was a member of AQIM and was arrested in 2007 and charged with terrorism for being affiliated with al-Qa'ida. He was sent to Abu Salim prison and was released just before the Libyan revolution in 2011. During the revolution, he fought in the battles in Sirte with Abdul-Fattah al-Basikri al-Mushaiti and his Salah al-Din Battalion. The battalion had a large number of terrorists who were charged with Taher in 2007, and members of the battalion were reported to have joined AAS-B when the battalion went defunct.

Benghazi: Know Thy Enemy

PERSONAL INFO
Full Name: Taher Mustafa Mohammad al-Awami
Affiliation: Al-Qa'ida in the Islamic Maghreb (AQIM)
Nationality: Libyan

BIOGRAPHY
In 2007, arrested and sent to Abu Salim prison for his role with AQIM

Freed prior to the 2011 Libyan Revolution and fought Gaddafi forces in Sirte, Libya in the extremist Salah al-Din Battalion led by Abdul-Fattah al-Mushaiti

On September 11, 2012, attacked the U.S. Consulate in Benghazi with AQIM

On February 10, 2014, Abdul-Fattah's family home exploded (see image); Abdul-Fattah, Taher and three additional AQIM associates inside to include Ahmed Mohammad al-Kawafi, Abdullah Sheikhi, and a third never identified terrorist were killed; The cause of the explosion was unknown

Taher al-Awami

U.S. Consulate Attacker
September 11, 2012

STATUS: DECEASED, KILLED IN A FEBRUARY 2014 EXPLOSION IN THE AL-LAYTHI NEIGHBORHOOD IN BENGHAZI

On September 11th, 2012, Taher attacked with U.S. Consulate several AQIM battalion members. He also attacked the Consulate with Abdul-Fattah's brothers and cousins to include Shoaib, Hamza, Salah, Ahmed, and Majdi al-Mushaiti.

On February 10th, 2014, the Mushaiti's family home in the al-Laythi neighborhood blew up. Inside the home were Abdul-Fattah, Taher, and three additional AQIM associates, including Ahmed Mohammad al-Kawafi, Abdullah Sheikhi, and a third never identified. After the explosion, all the bodies were transferred to the al-Jalaa Hospital in Benghazi. The unidentified AQIM member may still be preserved at the hospital. The cause of the explosion was unknown. The Mushaitis and several additional U.S. Consulate attackers were involved in the assassination of Major General Abdul Fatah Younis al-Obeidi, and it was believed this was in retaliation; however, these terrorists had a lot of local enemies after years of leading prolonged assassination campaigns.

(32) **Yousef al-Awami,** full name Yousef Idris Musa al-Awami from Benghazi. Yousef was the brother of fellow Benghazi attacker and facilitator Mahmoud al-Awami. He was a graduate of Garyounis University's

Department of Islamic Studies. In 2011, Yousef joined the Rafallah al-Sahati Brigade during the Libyan revolution. He was arrested by the Internal Security Forces at the beginning of the revolution but not held long. Then in 2012, he joined AAS-B and was one of the fighters for the group.

Yousef al-Awami

U.S. Consulate Attacker
September 11, 2012

STATUS: AT-LARGE, LOCATED IN EITHER TRIPOLI, LIBYA OR IN TURKEY; TWO ADDITIONAL BROTHERS RESIDED NEAR TRIPOLI AS WELL TO INCLUDE FELLOW ATTACKER MAHMOUD

PERSONAL INFO
Full Name: Yousef Idris Musa al-Awami
Affiliation: Rafallah al-Sahati Brigade, Ansar al-Sharia-Benghazi (AAS-B)
Nationality: Libyan

BIOGRAPHY
In 2011, during the Libyan Revolution, joined Rafallah al-Sahati; Arrested temporarily by Gaddafi's Internal Security Forces

In 2012, joined AAS-B; He was a fighter and his brother Mahmoud helped perform legal services

On September 10, 2012, he and his brother welcomed a number of foreign fighters, most from Egypt, to their family home; Attacked the U.S. Consulate with them the following day

Fearing U.S. counterstrikes, fled to Syria in 2012 with Benghazi attacker Hudhayfa bin Hariz

Injured in Syria, traveled to Turkey for medical treatment, and after treatment returned to Libya

On September 10th, 2012, he and his brother welcomed several foreign fighters to their family home. The terrorists had traveled to Benghazi to participate in the attacks on the U.S. Consulate. An interesting thing to note is that the brothers left home the evening of September 11th with the attackers, but when they returned after the attacks, none of their guests returned. Fearing U.S. counterstrikes, Yousef fled to Syria with Benghazi attacker Hudhayfa bin Hariz. He was injured in Syria and traveled to Turkey for medical treatment, and after medical treatment returned to Libya. As of 2022, it was unclear whether Yousef lived in Tripoli, Turkey, or both, as he frequently traveled between the two locations.

Mahmoud and Yousef also had a younger brother named Ahmed, who, like Yousef, had been a member of the Rafallah al-Sahati Brigade. As of 2022, Ahmed was located in Tripoli, fighting with the Government of Libya against the LNA. It was unknown if Ahmed participated in the U.S. Consulate attacks.

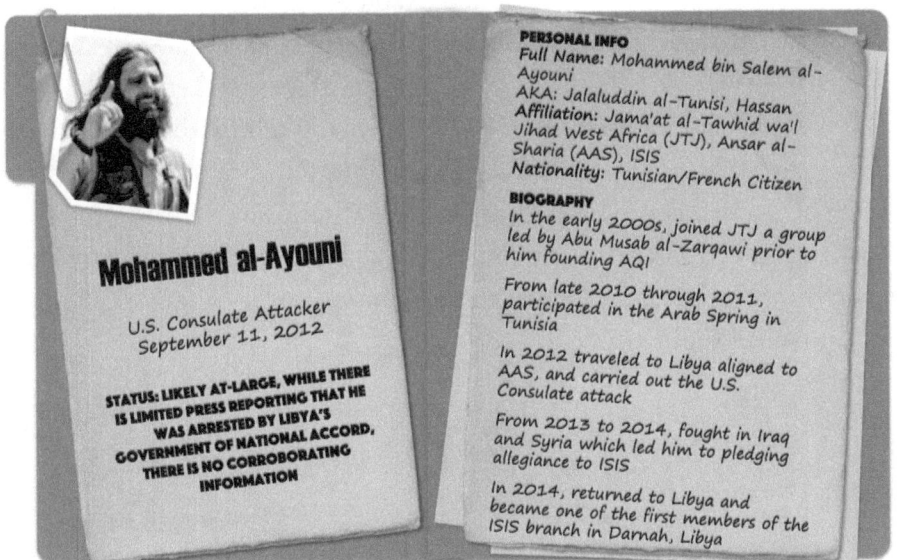

(33) Mohammed al-Ayouni, full name Mohammed bin Salem al-Ayouni with aliases Jalaluddin al-Tunisi and Hassan, from Tunisia and a former citizen of France. Historically, al-Ayouni was reported to be a member of Jama'at al-Tawhid wa'l Jihad (JTJ) West Africa, a group originally founded by AQI's Abu Musab al-Zarqawi. He participated in the Arab Spring uprising in Tunisia from late 2010 to the end of 2011. He then traveled to Libya in 2012 and was affiliated with members of Ansar al-Sharia in both Tunisia and Libya.

On September 11th, 2012, he participated in the attack on the U.S. Consulate in Benghazi. In the aftermath of the attacks, he fled to Syria. From 2013 to 2014, al-Ayouni fought in Iraq and Syria, which led him to join ISIS in 2014. Later in 2014, he returned to Libya when the Second Libyan War kicked off and after the LNA started counterterrorism operations in Benghazi. He was one of the original members of the ISIS Branch in Darnah. There was press reporting that Libya's Government of National Accord had arrested al-Ayouni; however, there was no corroborating information, and as of 2022, he was assessed to be at-large.

Of note, al-Ayouni's name, biography, and photograph get misconstrued with many terrorists. Most prominently, with two of the original ISIS Leaders in Libya, Abu Nabil al-Anbari and Abdul Qader al-Najdi. The other terrorist is Khaled Mohammed al-Saleh al-Ayouni, from the city of Nabeul, Tunisia, who held French citizenship and was a known terrorist in Syria. Khaled previously fought with AQI and was detained in Iraq. He was released from Iraqi custody and returned to Libya on September 1st, 2012, with the assistance of terrorist Hasan al-Salahayn Salih al-Sha'ari alias Abu Habib al-Libi who facilitated attackers to Benghazi for the U.S. Consulate attacks, including former AQI Senior Operative Ali Ouni al-Harzi.

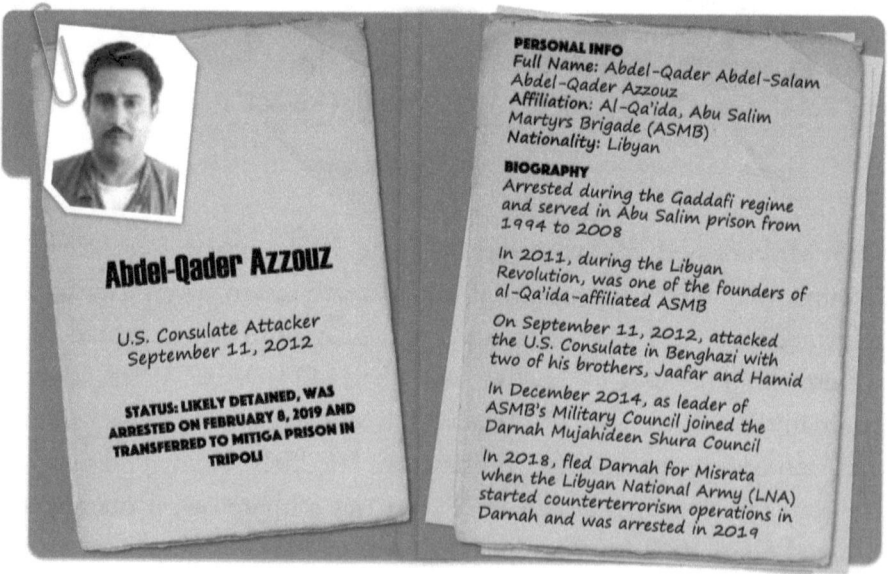

(34) Abdel-Qader Azzouz, full name Abdel-Qader Abdel-Salam Abdel-Qader Azzouz from the Al-Maghar region of Darnah, Libya. Azzouz was a member of al-Qa'ida. He was previously arrested by the Gaddafi regime and served in the Abu Salim prison from 1994 to 2008. In 2011, during the Libyan revolution, he was one of the founders of the al-Qa'ida-affiliated Abu Salim Martyrs Brigade (ASMB) based in Darnah. On September 11th, 2012, Azzouz attacked the U.S. Consulate in Benghazi with two of his brothers, Jafaar Azzouz and Hamid Azzouz.

After Abdel Hakim al-Hasadi left his position as the President of ASMB's Military Council, Azzouz became head the Council. In this position, he issued fatwas for the murder of military officers and judges. In April 2014, he admitted while on a local radio station in Darnah that ASMB was behind the killing of military officers and security officials and that the group was justified in killing judges as they were not ruling based on Sharia law. He also effectively used other broadcasting means like YouTube, expressing support for al-Qa'ida and Osama bin Laden.

In December 2014, the al-Qa'ida-affiliated DMSC was formed, and Azzouz and close associate and terrorist Magdy al-Houti became senior leaders. The council was to bring militias and terrorist groups together to counter General Haftar and the LNA and as an allied front against ISIS efforts to take control of Darnah. In addition to ASMB, Ansar al-Sharia-Darnah was a key member.

In 2018, Azzouz fled LNA targeting operations in Darnah for Misrata. He was arrested on February 8th, 2019, and transferred to Mitiga Prison in Tripoli.

(35) Jaafar Azzouz, full name Jaafar Abdel-Salam Abdel-Qader Azzouz from Darnah. Jaafar got his start in terrorism in the EIJ. Jaafar accumulated several aliases over the years, including Sharhabeel al-Andalus, Yazid Abu Khathaima, Al-Tarid al-Sahi, Al-Ghafir Adam al-Sudani, Marja' Bungath, and Al-Barqawi. In 2007, he was detained for supporting global terrorist networks including al-Qa'ida. Jaafar was a member of al-Qa'ida in the Islamic Maghreb's (AQIM) Mali Group. He was also the brother of Abdel-Qader Azzouz, one of the co-founders of ASMB.

On September 11th, 2012, Jaafar attacked the Consulate in Benghazi with his two brothers Abdel-Qader and Hamid. Also participating in the Consulate attack with the Azzouz brothers were several AQIM Mali Group members, including Hashem Bousidra, Mahdi Dango, Abu Hamza al-Tabawi, and Mansour al-Shalaali (the brother of Omar al-Shalaali).

In 2014, Jaafar was one of the founders of the Islamic Army in Darnah. He was also one of the founders of the Emirate of Borders and Immigration, affiliated with ISIS, in Sirte. On February 12th, 2015, he, along with terrorist Walid al-Ferjani and Benghazi attackers Hashem Bousidra, and Mahdi Dango were involved in kidnapping 21 Egyptian Coptic Christians and then massacred them on February 15th, 2015. On February 16th, the Egyptian Military responded with airstrikes against ISIS, and the LNA followed suit with a strike at the Al-Jabal Company in Darnah.

On January 27th, 2017, Jaafar was reportedly killed in a U.S. dual airstrike north of Sirte. This strike killed 80 ISIS terrorists at two separate training camps. While Jaafar was likely killed in the Sirte strike, there was reporting that Jaafar may still be at-large and in hiding in Jabal al-Shaabni, Tunisia. Separately, regarding his terrorist brothers, on February 8th, 2019, Abdel-Qader was detained, and on February 21st, 2016, Hamid was reportedly killed by the al-Qa'ida-affiliated DMSC.

B

(36) Zubayr al-Bakoush, full name Zubayr Hassan Omar al-Bakoush from Benghazi. Before the Libyan revolution in 2011, Zubayr was not reported to have been involved in terrorism and was known to have spent years in the Scouts. The Boy Scouts started in Libya back in the 1950s. Unfortunately, over the years, terrorist groups in Libya had attempted to find youths to recruit from both the Scouts and the Libyan Red Crescent youth teams.

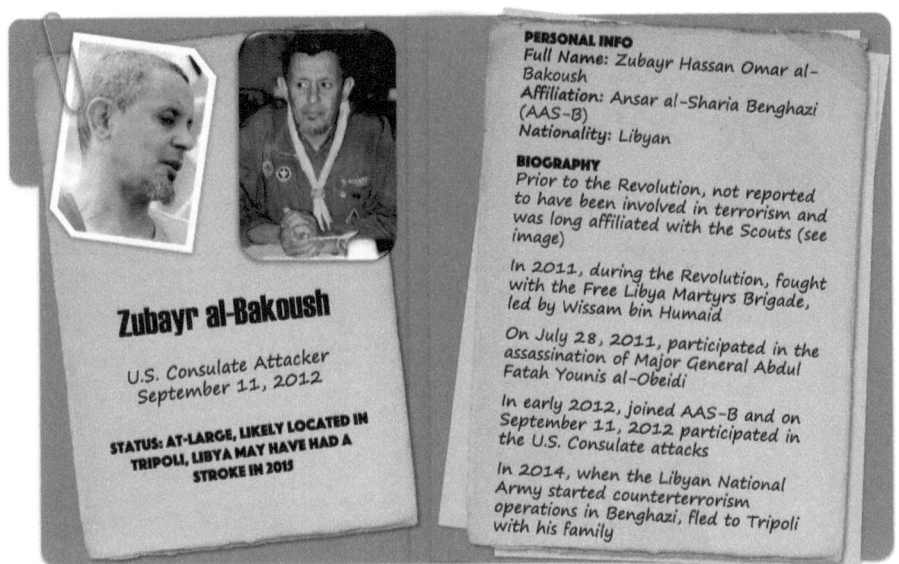

Zubayr al-Bakoush
U.S. Consulate Attacker
September 11, 2012

STATUS: AT-LARGE, LIKELY LOCATED IN TRIPOLI, LIBYA MAY HAVE HAD A STROKE IN 2015

PERSONAL INFO
Full Name: Zubayr Hassan Omar al-Bakoush
Affiliation: Ansar al-Sharia Benghazi (AAS-B)
Nationality: Libyan

BIOGRAPHY
Prior to the Revolution, not reported to have been involved in terrorism and was long affiliated with the Scouts (see image)

In 2011, during the Revolution, fought with the Free Libya Martyrs Brigade, led by Wissam bin Humaid

On July 28, 2011, participated in the assassination of Major General Abdul Fatah Younis al-Obeidi

In early 2012, joined AAS-B and on September 11, 2012 participated in the U.S. Consulate attacks

In 2014, when the Libyan National Army started counterterrorism operations in Benghazi, fled to Tripoli with his family

Zubayr fought with the Free Libya Martyrs Brigade during the Libyan revolution, led by Wissam bin Humaid. On July 28th, 2011, he participated in the assassination of Major General Abdul Fatah Younis al-Obeidi. In early 2012, he joined AAS-B and carried out the attacks on the U.S. Consulate in Benghazi as a member of the group. In 2014, when the LNA started counterterrorism operations in Benghazi, Zubayr fled to Tripoli with his family. As of 2022, he was still assessed to be located in Tripoli.

(37) Ziad Balaam, Farid Mohammad Mohammad Balaam with alias Omar al-Mokhtar from Benghazi. In 2002, Ziad was sentenced to life in Abu Salim prison for being a member of al-Qa'ida after being involved in terrorist activities in Sudan. In 2011, during the revolution, he was first affiliated with the Omar Mokhtar Brigade out of Ajdabiya, Libya. Ziad then linked up with a network of al-Qa'ida affiliates from Iran to fight in the revolution, including Abdel Moneim al-Madhouni. Madhouni was the founder of al-Qa'ida's Malik Brigade variant Saraya Malik which reported directly to al-Qa'ida Senior Leadership (AQSL). Madhouni was killed on April 15th, 2015, in the vicinity of al-Burayqah, Libya.

Enemies: Brothers in Arms—The Attackers

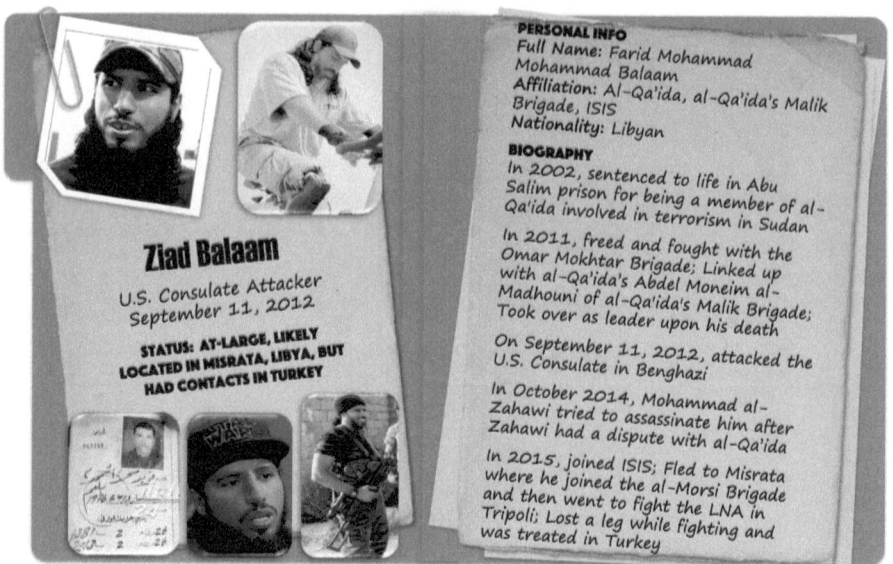

Madhouni was the LIFG's senior shura council leader and went to fight in Afghanistan with Osama bin Laden and al-Qa'ida. After 9/11, he fled to Iran and did not return to Libya until the revolution. When he died, there was a ceremony videotaped of him at the same hospital in Benghazi, where good Samaritans brought our Ambassador in the early hours of September 12th, 2012. The ceremony had several al-Qa'ida members, and then Benghazi attacker Mohammad al-Kawil spoke about Madhouni and his roots in Sabratha. After Madhouni's death, Ziad became the Leader of the al-Qa'ida's Malik Brigade.

On September 11th, 2012, Ziad attacked the U.S. Consulate in Benghazi and claimed the Squad Automatic Weapon (SAW) he still used a decade later was stolen during the attacks. In 2014, after the start of the Battle of Benghazi, a legal dispute occurred between al-Qa'ida and AAS-B. In the dispute, al-Qa'ida first wanted to take over all of Libya and then declare it an Islamic caliphate. On the other hand, AAS-B wanted to declare Benghazi the Islamic caliphate present day, and then as each region was taken over, they would be rolled into the caliphate. AAS-B had already wanted to declare the caliphate as it was the main impetus behind the group's formation in 2012. Hence, the group was impatient to get the ball rolling, while al-Qa'ida as an organization was working on a longer, more strategic timeline in their planning. Ziad was the al-Qa'ida representative to handle the dispute, and Mohammad al-Zahawi was the AAS-B representative.

Instead of solving the issue in vision between the groups in a civil manner, Zahawi went full jihadi and sent an assassin to kill Ziad. The assassin was one of the al-Rubaie brothers and most likely was Benghazi attacker Ramadan al-Rubaie. Ziad was injured severely in the assassination attempt and had to travel to Turkey for medical treatment. As a result of this incident, the Malik Brigade withdrew from Benghazi. It relocated to the Seddah Bridge, where extremist groups had a camp called the Citadel, located in the region between the cities of Bani Walid and Misrata.

In April 2019, when the LNA started counterterrorism operations in Tripoli, the Malik Brigade joined forces and fought with Fayez Al-Sarraj's Government of National Accord (GNA) and its Morsi Brigade against the LNA. In the first few days of the battle in Tripoli, Ziad was filmed where he said he was there fighting to avenge the death of CIA Annex attack Mastermind Wissam Bin Humaid. While battling the LNA, in March 2020, Ziad was shot, severely injured, and again taken to Turkey for medical treatment. He had his leg amputated as a result of his injuries. He arrived back in Libya in early August 2020 and kept a low profile until June 5th, 2021, when he showed up at a Misrata militia's graduation ceremony.

As of 2022, Ziad was at-large and residing between the cities of Misrata and Tripoli. He also had freedom of movement in Turkey. He reportedly also had a relationship with Turkish Intelligence as he had sent terrorists to fight in Syria in support of the Turkish Government.

(38) Mahmoud al-Barassi, full name Mahmoud Ali al-Barassi, with aliases Abu Musab al-Libi, Abu Musab Houda al-Baghdadi and previously known as Abu Musab al-Farouq, a resident of the al-Sabri neighborhood in Benghazi, was a former member of AAS-B's Shura Council at the time of the attacks. During the Libyan revolution in 2011, he was a member of the Rafallah al-Sahati Brigade.

Enemies: Brothers in Arms—The Attackers

Mahmoud al-Barassi
U.S. Consulate Attacker
September 11, 2012

STATUS: LIKELY DECEASED, KILLED ON SEPTEMBER 27, 2019 DURING A U.S. STRIKE TARGETING ISIS IN SABHA, LIBYA; AS NO PHOTOGRAPHIC EVIDENCE WAS AVAILABLE OF HIS DEATH, SOME BELIEVE HE MAY STILL BE AT-LARGE DUE TO HIS WEALTH

PERSONAL INFO
Full Name: Mahmoud Masoud al-Barassi
AKA: Abu Musab al-Libi, Abu Musab al-Farouq, Houda
Affiliation: Ansar al-Sharia-Benghazi (AAS-B), al-Qa'ida, ISIS
Nationality: Libya

BIOGRAPHY
In 2012, was a member AAS-B's Shura Council and was leading assassination campaigns in the city of Benghazi against government officials

In 2013, ran an al-Qa'ida terrorist training camp in Darnah, Libya, and ended up breaking with AAS-B after making public threats against Libya's General National Congress and Army

In 2014, swore allegiance to ISIS Leader Abu Bakr al-Baghdadi and founded the ISIS Branch in Benghazi

By 2017, was a senior leader of ISIS remnants in both Ajdabiya and Bani Walid, Libya

In 2012, Barassi was well-known for being involved in targeted assassinations against the government, military, and security officials in Benghazi, particularly in the al-Sabri neighborhood. He terrorized the city of Benghazi in the months leading up to the Consulate and Annex attacks. At the time of the attacks, he was one of the most extreme terrorists in the city and was later known for operating an al-Qa'ida terrorist training camp in Darnah.

While in Benghazi, and still as a member of AAS-B, Barassi was operating a network of terrorists made up of former AQI members, including Ahmed Abd al-Salam al-Hami with alias Abu Anas. Hami had traveled from Iraq with several Iraqi detainee terrorists who were released from Iraqi custody due to a request from the Government of Libya. Additional terrorists in Barassi's local network at the time included: Ayman Mohammad Salih al-Rajhi, Sassi Salem Sassi Mohammad with alias Abu Mohammad al-Misrati, and Walid Hussein Salem Burayaqah with alias al-Qaqaa. These network members were eventually arrested and held by the Special Deterrence Force in Tripoli.

Barassi eventually broke from AAS-B over a dispute with its Leader Mohammad al-Zahawi over declaring the Emirate of North Africa in

Benghazi. Barassi believed Zahawi was waiting too long to announce this. During this spat, Barassi operated primarily in Darnah and appeared on national television denouncing the Libyan General National Congress (GNC) and the country's Libyan Army calling them "apostates." This denouncement meant these government organizations had abandoned the fundamental beliefs of their religion of Islam.

In 2014, Barassi left Darnah and returned to Benghazi. There he took on a senior role as the founder of the ISIS branch in Benghazi after swearing allegiance to ISIS Leader Abu Bakr al-Baghdadi. Barassi took some key terrorists with him when he defected from AAS-B, including Abdullah Bala, Omar al-Tunisi, and Benghazi attackers Ahmed Bin Nasser and Ahmed al-Mushaiti.

During the Second Libyan Civil War, Barassi led ISIS assaults on the LNA in the northern axis in Benghazi, this included areas surrounding the al-Hout Market, the al-Sabri neighborhood, the Municipal Hotel, and the Gold Market. The front led by Barassi collapsed due to an insider within his group that provided actionable intelligence to the LNA. Barassi escaped, but many of his men were captured and killed due to this insider incident. On his way out of Benghazi, Barassi fled with millions of dollars stolen from banks in Benghazi and gold holdings heisted from the Gold Market.

Barassi was first believed to have fled to desert areas outside Sirte, Libya. Then after ISIS fell in Sirte in December 2016, just a little over five years after Gaddafi fell in Sirte, he was one of a handful of ISIS Leaders still commanding the group. He went on to lead pockets of ISIS fighters operating in Ajdabiya, south of Benghazi, and Bani Walid, southeast of Tripoli and southwest of Misrata. At the time, he was involved in attacking remote police stations.

It was then reported that Barassi ditched his network and headed farther into western Libya, where he sought refuge with terrorist Shabaan Hadiyah al-Makani with alias Abu Ubaydah al-Zawi. Shabaan led the Libyan Revolutionaries Operations Room. He had returned from Yemen and was wanted in international terrorism cases. As of 2022, the militia run by Shabaan was still active in western Libya and was operating in Tajoura or the Friday market area, where most of the terrorists of the al-Qa'ida-affiliated Shura Councils of Benghazi, Darnah, Ajdabiya, and Sirte relocated to after the LNA pushed them out of eastern Libya.

Barassi was first rumored to have died at an ISIS headquarters meeting that was targeted in 2016. Then just months after filming a video encouraging an ISIS resurgence in Libya, he was reportedly killed on September 27th, 2019, during a U.S. strike targeting a large gathering of ISIS members in Sabha, in southern Libya.

While he was likely deceased, since there is no photographic evidence of his death, we note that there is an alternate storyline where Barassi was severely injured in the same strike. He may have gone to the Rahba al-Duru camp in Tajoura, Libya, where dozens of fighters fleeing from Benghazi were taken care of by terrorist Bashir al-Baqara. Barassi successfully kept his identity a secret over the years, so he would not have been easily identifiable in the camp. Even the Public Prosecutor in Tripoli at the time, Al-Siddiq Al-Sour, had the wrong photograph published for Barassi. These errors could have allowed him to escape easily. Ironically, when BBC filmed AAS-B's Leader Zahawi right after the 2012 Benghazi attacks, they captured Barassi on video, but no one realized it.

Besides being able to travel unnoticed, as most in Libya knew his name but had no idea what he looked like, Barassi again was a wealthy terrorist. He had been providing funds to his brother Buajila al-Barassi and another terrorist named Allawi Bushnaf, who had been residing in Turkey for years. And it was believed that Barassi may have traveled to Turkey and lived under a false identity as he had the means to do so.

(39) Marwan al-Barki, full name Marwan Hassan al-Barki with alias Abu Darda from the al-Laythi neighborhood in Benghazi. Marwan joined al-Qa'ida in 2003. He was also a mentee of terrorist Salem Saleh al-Fakhri in Ajdabiya, Libya, at the time. After being educated in jihadist principles, he became the Imam of the Al-Noor al-Mubin Mosque in the al-Laythi neighborhood. This mosque was involved in indoctrinating terrorists.

In 2011, he fought in the Libyan revolution with Ahmed Abu Khatallah and his neighborhood battalion, the Ubaydah bin Jarrah (UBJ) Battalion. At the time, Marwan was a security escort for Khatallah. In

2012, Marwan joined AAS-B and participated as a group member in the September 11th attack on the U.S. Consulate in Benghazi. Starting in May 2014, he fought in the Battle of Benghazi against LNA.

Later in 2014, former AAS-B Senior Commander and Benghazi attacker Mahmoud al-Barassi recruited Marwan to ISIS in Benghazi. He was primarily focused on attacking LNA and other security-related checkpoints in the city. He ended up fleeing LNA counterterrorism operations for Ajdabiya. On June 2nd, 2018, he participated in an ISIS attack on a police station in al-Qanan, just east of Ajdabiya. ISIS went deeper into the desert as LNA operations closed in on them in Ajdabiya, and information on Marwan went cold. As of 2022, his current status and whereabouts were unknown.

(40) Omar al-Barki, full name Omar Mohamed Zayed al-Barki with alias Tamtam from Benghazi. In 2004, he joined AQI and was captured in Syria while traveling to fight U.S. forces in Iraq. His family reported him missing, and he ended up being deported from Syria back to Libya. Upon his return, he was tried, sentenced, and sent to Abu Salim prison for his involvement with AQI. In 2011, during the Libyan revolution,

Omar joined 17 February. He fought with the group in Sirte, against Gaddafi's forces, where he became a senior leader.

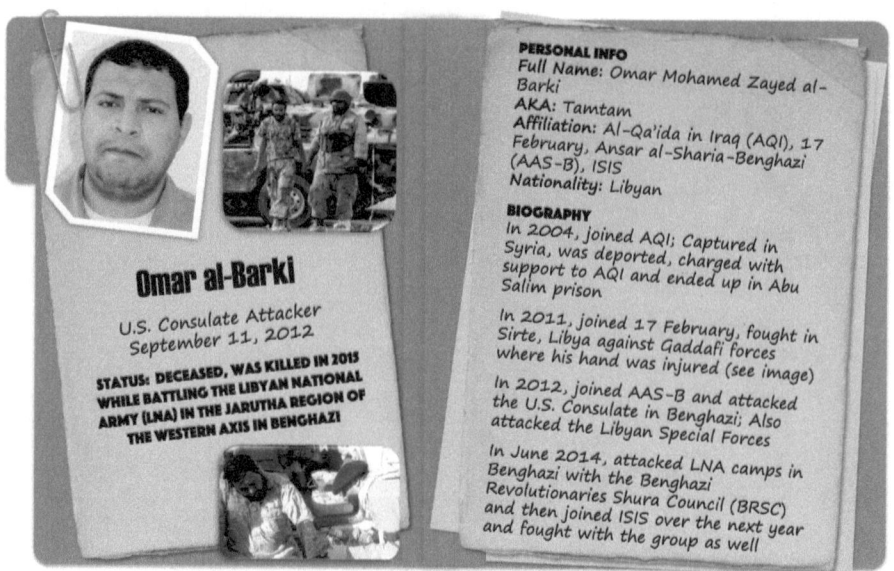

On September 11th, 2012, he participated in the attacks on the U.S. Consulate in Benghazi. He was later involved on September 27th, 2012, in the AAS-B attacks on the Benghazi Council of the Benghazi Security Directorate, where the Lightning Battalion (Libyan Special Forces, also known as the Saiqa Brigade) was based.

In June 2014, Omar participated in the attack on the LNA camps. This incident was significant because it allied most terrorist groups in the city as they rallied under the BRSC banner. These battles also eventually pushed al-Qa'ida and ISIS terrorists together with LNA as their common enemy. The groups would subsequently split apart, but it is why you see so many attackers fight in Benghazi with ISIS in the 2014 and 2015 timeframe. Omar fought with ISIS, as well. In 2015, he was killed while battling the LNA in Benghazi's Jarutha region of the western axis.

Omar was the brother of Saleh Omran al-Barki. He led the Abu Salim Military Council and fought for the Government of Libya as a Senior Commander in Libya Dawn against the LNA. Saleh was also killed in 2015 by the LNA.

(41) **Ayman Bouamoud,** full name Ayman Faraj Hassan Bouamoud al-Bargathi with alias Miqdad and Zawiya from Benghazi. Ayman was a Hamza Bouchertila's group member and then joined the Martyrs Group with Hamza. Ayman also became a member of al-Qa'ida after the Arab Spring. In 2004, he was arrested for supporting the facilitation of terrorists in Algeria and Iraq. He was released in 2010 and received compensation from the Libyan Government for being detained.

In 2011, during the Libyan revolution, he joined 17 February. In late 2011, nearing the end of the revolution, he joined al-Qa'ida's Malik Brigade. Again, the Brigade was led in 2012 by Abdel Moneim al-Madhouni until he died in 2015, and Benghazi attacker Ziad Balaam took over.

In 2012, Ayman was involved in many terrorist operations in Benghazi, including assassinations, kidnappings, and bombings, against military and security personnel. On September 11th, 2012, Ayman attacked the U.S. Consulate in Benghazi, likely with members of his Brigade, including Ziad Balaam. He also attacked the Consulate with his cousin, Zakaria Bil Qasim Harroun al-Bargathi, with alias Jutuf. Zakaria was the terrorist that called Khatallah eight minutes into the attack to notify him that an attack was occurring at the Consulate.

Enemies: Brothers in Arms—The Attackers

On April 15th, 2014, Ayman kidnapped Jordanian Ambassador Fawaz al-Aitan with the brothers of detained Libyan terrorist Mohamed el-Dresi. The mastermind of the attack was Dresi's brother Ahmed Saeed al-Nass. Dresi was a member of AQI and was released in exchange for the Ambassador. The Ambassador was released on May 13th, 2014.

On October 21st, 2014, Ayman was arrested in the al-Humaydah area of Benghazi by the Saiqa Brigade. Dresi repaid the favor in November 2014 and carried out the kidnapping of three of Colonel Salah Buhaliqa al-Urfi's brothers to exchange for Ayman. Ayman was released in exchange for four of the Colonel's brothers (Khaled, Shukri, Alaa, and Adel) as terrorists had already detained one brother since August 2014. Also in the exchange negotiations for Ayman was Abdel Salam Khaled, the nephew of Major General Jamal al-Zahawi. On January 21st, 2015, Ayman was released in exchange, Dresi was there, and one of the Shalaali brothers was there. This was likely Omar al-Shalaali, as he was the al-Qa'ida's Leader for East Libya at the time. As a reminder, Omar was the commander who led the attacks at the Consulate on September 11th.

In 2014, Ayman joined the BRSC. In 2015, during the Battle of Benghazi against the LNA, Ayman fled to Misrata. For his security, Ayman then rotated through the cities of Misrata, Zliten, Tajoura, Tripoli, and al-Zawiya. In June 2016, he then joined the Benghazi Defense Brigades.

In March 2017, Ayman was injured during clashes in Tripoli. On January 15th, 2018, he was involved in the attack on the Mitiga airport and prison in Tripoli. The attack was likely to break out terrorist allies, which may have included several Benghazi attackers that were held there. The other terrorists who participated in this attack included Misbah al-Sheikhi, Bashir Khalaf Allah with alias al-Baqara, Al-Mahdi al-Kilani, Khaled al-Qalal, and Sharif Boulifa. Sharif during the revolution had been a Leader in the Free Libya Martyrs Brigade, which was led by Annex attack Mastermind Wissam bin Hamid and Annex attack suspect Hafez al-Aqouri.

At some point in 2018, Ayman also traveled to Turkey to receive medical treatment. As of 2022, he was at-large and assessed to be located in western Libya. His Benghazi attacker cousin Zakaria was also still at-large.

Benghazi: Know Thy Enemy

PERSONAL INFO
Full Name: Walid Abdel Qader Mohammed al-Barnawi
Affiliation: Ubaydah bin Jarrah (UBJ) Battalion, Ansar al-Sharia-Benghazi (AAS-B)
Nationality: Libyan

BIOGRAPHY
In 2011, during the Libyan Revolution, joined Ahmed Abu Khatallah's neighborhood militia UBJ

In 2012, joined AAS-B and participated in the attacks on the U.S. Consulate in Benghazi; Led assassinations against security officials in the al-Laythi area of Benghazi

In 2014, was a AAS-B Commander for al-Laythi where he fought the Libyan National Army (LNA); In 2017, fled Benghazi after LNA liberated the city from terrorists

He relocated to Tripoli and was arrested by the RADA Special Deterrence Forces, and as of 2022 was likely still detained

Walid al-Barnawi

U.S. Consulate Attacker
September 11, 2012

STATUS: DETAINED, LIKELY LOCATED AT THE MITIGA PRISON IN TRIPOLI

(42) **Walid al-Barnawi,** full name Walid Abdel Qader Mohammed al-Barnawi from the al-Laythi neighborhood in Benghazi. Before the Libyan revolution, he was not known to have been involved in terrorism. In 2011, he joined Khatallah's neighborhood militia Ubaydah bin Jarrah (UBJ) Battalion. In 2012, he joined AAS-B, and on September 11th, 2012, he participated in the attacks on the U.S. Consulate in Benghazi. Walid led assassinations against the police, military, and security officials, primarily in the al-Laythi area. He had extreme views and hostility primarily against the LNA, the key target for his terrorist operations.

During the Battle for Benghazi in 2014, Walid was an AAS-B Commander fighting the LNA in the al-Laythi axis. Walid's home was used as an operations room for AAS-B, and weapons and ammunition were stored at the location. In March 2015, he may have been injured fighting with the BRSC against the LNA and may have traveled to Misrata at the time for medical treatment. In approximately 2017, after LNA counterterrorism operations defeated the terrorist base in Benghazi, Walid fled to Tripoli. In Tripoli, he was arrested by the RADA Special Deterrence Forces. RADA was an Islamist special operations military police unit formed by Libya's Ministry of Interior (MOI) in Tripoli,

Libya. As of 2022, he was likely located at the Mitiga Prison in Tripoli.

(43) Salim Bayou, full name Salim Mustafa Nasser Bayou from Benghazi. Salim was a former prisoner of Abu Salim prison. In 2007, Salim was detained in the same counterterrorism case involving at least eighteen other Benghazi attackers for supporting global terrorist networks to include al-Qa'ida. These terrorists when they were detained hailed from several extremist and terrorist groups based in Benghazi, Darnah, and Ajdabiya. Less than a handful of those released returned to a normal productive life. The rest were involved in terrorism or died from being involved. It is important to understand that the recidivism rate from Abu Salim prison is astronomically high compared to the Guantanamo recidivists' figures.

In 2012, Salim joined AAS-B and was also featured in an AAS-B publication. He led sermons and not only incited violence against the army, the police, and state institutions, but he also participated in several crimes and terrorist activities in Benghazi. Salim's primary role with AAS-B was appalling as he was charged with recruiting children to be members of the terrorist group and, as time went on, to other terrorist groups like al-Qa'ida and ISIS. For example, on September 11th, 2012, when he attacked the U.S. Consulate, he had one of his recruits with him, Benghazi attacker

Hudhayfa bin Hariz, who was just a teen. Salim was responsible for several children being captured or killed in Iraq and Syria. After the Benghazi attacks, Salim traveled to fight in Syria for the al-Nusrah front.

Salim left Syria in 2013 to return to Benghazi. He fought against the LNA during its Operation Dignity in the Battle for Benghazi. In September 2014, Salim was killed in battle, likely by the LNA.

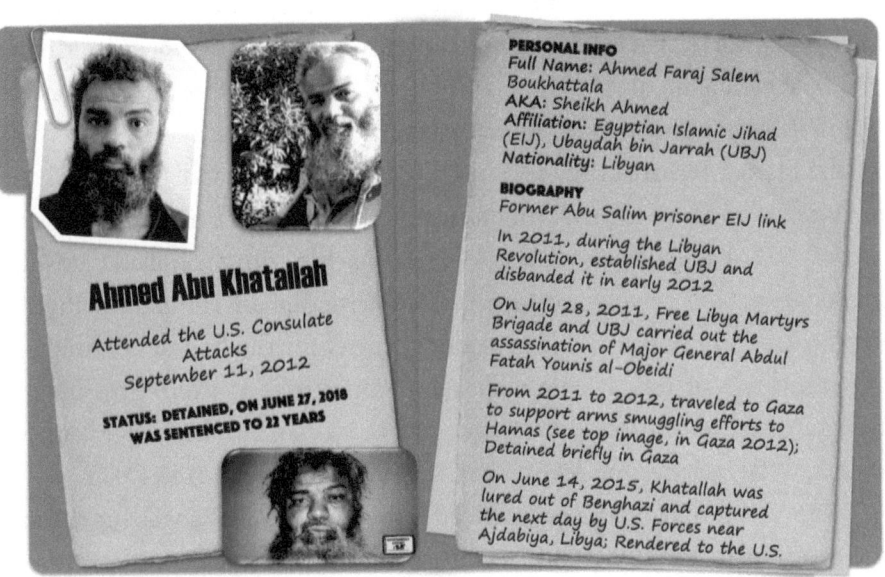

Ahmed Abu Khatallah
Attended the U.S. Consulate Attacks
September 11, 2012
STATUS: DETAINED, ON JUNE 27, 2018 WAS SENTENCED TO 22 YEARS

PERSONAL INFO
Full Name: Ahmed Faraj Salem Boukhattala
AKA: Sheikh Ahmed
Affiliation: Egyptian Islamic Jihad (EIJ), Ubaydah bin Jarrah (UBJ)
Nationality: Libyan

BIOGRAPHY
Former Abu Salim prisoner EIJ link

In 2011, during the Libyan Revolution, established UBJ and disbanded it in early 2012

On July 28, 2011, Free Libya Martyrs Brigade and UBJ carried out the assassination of Major General Abdul Fatah Younis al-Obeidi

From 2011 to 2012, traveled to Gaza to support arms smuggling efforts to Hamas (see top image, in Gaza 2012); Detained briefly in Gaza

On June 14, 2015, Khatallah was lured out of Benghazi and captured the next day by U.S. Forces near Ajdabiya, Libya; Rendered to the U.S.

(44) **Ahmed Abu Khatallah,** full name Ahmed Faraj Salem Boukhattala with alias Sheikh Ahmed (even though he had no formal religious training) from Benghazi. Khatallah was a former prisoner in Abu Salim prison and was detained due to ties to the EIJ. He reportedly was not detained long, released, and placed under surveillance. His day job was as a contractor, and he was a professional home builder.

In 2011, during the Libyan revolution, he created Ubaydah bin Jarrah (UBJ), a small militia of approximately 20 terrorists from his neighborhood in Benghazi. Khatallah was from al-Laythi, a neighborhood so extreme that it was referred to locally as Kandahar, Libya. Khatallah was involved in several assassinations in the city, and he focused efforts on internal security officers. While assassinating citizens trying to establish a new government and a non-militia security apparatus in Benghazi,

Khatallah received funding from the Libyan Government in Tripoli to fund his militia activities.

On July 28th, 2011, a grouping of militia members from the Free Libya Martyrs Brigade and UBJ assassinated Major General Abdul Fatah Younis al-Obeidi. Wissam bin Humaid led Free Libya Martyrs Brigade, and his brother Mohammad bin Humaid participated in the assassination with Khatallah. After the assassination, it elevated Khatallah's popularity in Benghazi. However, he was no strategist and was known to be simple-minded. Khatallah and Wissam bin Humaid were also in the Hamas Cell together, where they would smuggle weapons to Egypt and onward to Gaza for the Hamas terrorist organization. During one of Khatallah's several trips to Gaza in the 2011 to 2012 timeframe, he was arrested briefly and released.

In 2011, Khatallah also had a close relationship with the Italian Consulate in Benghazi due to relationships created with Italian companies through his construction business. He provided security for the Italian medical staff, the employees of Italian companies, and diplomats at their Consulate in Benghazi. After the revolution, Khatallah disbanded his neighborhood militia, UBJ, with some former members joining AAS-B in 2012.

Reporting that UBJ was involved in September 2012 was incorrect as the group was defunct well before the attacks. There were several former members of UBJ involved in the attacks. However, terrorists from defunct militias were at the attacks under direction of either new group affiliations like AAS-B or old ones like al-Qa'ida. Most terrorists joined militias to fight in the revolution, but if they were AQIM, they stayed AQIM. Most militias were a temporary fighting force. Even when AAS-B was formed, it was an umbrella organization. You could still stay with AQIM and be a member of AAS-B, with Benghazi attacker Mansour al-Shalaali being a good example. Or, if you had no prior terrorist group affiliations, you could be solely a member of AAS-B.

On September 11th, 2012, Khatallah and Mustafa al-Imam, found out about the attacks right after they started. Khatallah had been at home having tea with a friend when he received the first call. He and Mustafa then spoke and made plans to meet outside the Consulate, where they loitered outside the complex. They went inside with a now-deceased

AAS-B Commander Khalid Nayhum after all the al-Qa'ida attackers and American personnel had fled and were involved in some of the looting inside the grounds. While it was reported that Khatallah stole items of value from the Consulate, it is important to note that the Consulate maintained no classified holdings.

Early on September 12th, 2012, Ambassador Stevens was found in his Villa and rescued by good Samaritans. Khatallah took it upon himself to capture one of the good Samaritans, a young Libyan Army officer who showed up after the attacks to look at the aftermath of the events. Khatallah, for hours, questioned everyone in the vicinity of the Consulate, including looking at phones from eyewitnesses. He was also asking basic questions about who led and carried out the attacks, showing a disconnect from prior planning efforts al-Qa'ida had with local terrorists.

Al-Qa'ida, though, would not need to have contacted Khatallah; as he had no prior relationship with the group and was leading no militia or group at the time. The fact he had no prior relationship to al-Qa'ida may have been the reason he was set up as the Mastermind as there seemed to be efforts within our own government to cover up al-Qa'ida's direct involvement in the attacks. After releasing the good Samaritan, Khatallah went home. He never traveled near the vicinity of the CIA Annex in Benghazi that evening and is not believed to have played a role in the attacks on the Annex either.

In the aftermath of the attacks, Khatallah could be found most days working on his various construction sites around the city. He never went into hiding and even held interviews with international press personalities, including Steven Sotloff. Khatallah was lured out of Benghazi and captured June 15th, 2014 by U.S. Special Forces near Ajdabiya, Libya.

On November 28th, 2017, a jury in Washington acquitted Abu Khatallah of 14 of the 18 charges he faced. On June 27th, 2018, Abu Khatallah was sentenced to 22 years. In the U.S. Justice system, he was essentially given 22 years for looting the U.S. Consulate in Benghazi. Khatallah though had many crimes to atone for and should be put away forever for his assassinations against brave Libyans. These Libyans wanted to serve their new nation in the police, army, and security services, and they were killed for wanting to do what's right.

Enemies: Brothers in Arms—The Attackers

Abdullah Bouzkia

U.S. Consulate Attacker
September 11, 2012

STATUS: LIKELY DETAINED, AS OF 2022, WAS ASSESSED TO BE IMPRISONED AT THE MITIGA PRISON IN TRIPOLI, LIBYA

PERSONAL INFO
Full Name: Abdullah Mohamed Mohamed Bouzkia
Affiliation: Egyptian Islamic Jihad (EIJ), Ansar al-Sharia–Benghazi (AAS-B)
Nationality: Libyan

BIOGRAPHY
In 1993, captured at the Libyan border, sentenced to life in Abu Salim prison for being affiliated with EIJ; Freed from prison just prior to the Libyan Revolution in 2011

In 2012, joined AAS-B, attacked the the U.S. Consulate in Benghazi; Led AAS-B assassination squads

In 2014, joined Benghazi Revolutionary Shura Council (BRSC) and was based at Qawarsha Gate during the Battle for Benghazi against the Libyan National Army (LNA)

Was reported to have been killed fighting in Benghazi, but likely actually fled to Tripoli and was detained by RADA Special Deterrence Forces

(45) Abdullah Bouzkia, full name Abdullah Mohamed Mohamed Bouzkia from the al-Muhajireen neighborhood in Benghazi. Bouzkia was involved in terrorism back to at least the early 1990s. In 1993, he was captured at the Libyan border after an armed group he was supporting clashed with the Sudanese Army, with members of his group killed. Sudan deported him to Libya, where he was tried and sentenced to life in prison. During the trial, he admitted to being affiliated with the Jihad Organization and the Mujahideen Brigades, which was affiliated with the EIJ and followed the guidance of Dr. Ayman al-Zawahiri at the time. He confessed that terrorist Jamal al-Zawi recruited him. He was sent to Abu Salim prison and later released during the Arab Spring as he faked a health condition, pretending to need a wheelchair.

In 2012, Bouzkia joined AAS-B and led assassination squads for AAS-B's Leader Zahawi. Then in 2014, he joined BRSC. In Benghazi, Bouzkia was based at the infamous Qawarsha Gate. At these terrorist headquarters, the following Benghazi attackers were based: Fawzi al-Faydi, Talal bin Hariz, Ahmed al-Mushaiti, and Mohammad al-Manfi. Further, terrorist Salem Shatwan was also based at the Qawarsha Gate.

Bouzkia also formed a militia that took control of the Bouhdima military prison in the fall of 2014. Many Darnah-based terrorists were sent to this prison as the LNA moved on from Benghazi to remove the terrorist base in Darnah. He fought the LNA in Benghazi during the Second Libyan Civil War and was initially reported to have been killed. However, he likely had fled Benghazi and was in hiding shifting between Misrata and Tripoli. While in Tripoli, he was reportedly captured by the RADA Special Deterrence Force. As of 2022, Bouzkia was believed to be a detainee imprisoned in Mitiga Prison in Tripoli.

(46) Ahmed Buhajar, full name Ahmed Mukhtar Buhajar with alias Hajari, from Benghazi. In 2011, during the Libyan revolution, he was an operational leader in Wissam bin Humaid's militia, Free Libya Martyrs Brigade. Before the revolution, he was a tradesman working in a family business. His involvement with the Free Libya Martyrs Brigade inspired him to commit terrorism. When he participated in the Benghazi attacks on September 11th, 2012, Ahmed was a member of Libya Shield One under Wissam.

After participating in the Benghazi attacks and fearing reappraisals from the U.S. Government, Ahmed fled with a number of the attackers to Syria. Ahmed joined and fought with the al-Nusrah Front in Syria and was wounded in the eye while fighting. He returned to Libya in 2014 during Operation Dignity, which started on May 16th, 2014. Ahmed fought with AAS-B terrorists against LNA. During the events in 2014, al-Qa'ida-affiliated terrorist groups AAS-B and the BRSC were able to amass a force of 4,000 strong to fight in Benghazi.

Ahmed was filmed while fighting in the Battle of Benina Airport during the Second Libyan Civil War, which lasted from August 2014 until October 2014. In the video, he states (from Arabic translated to English): "The battle of Libya is the battle of the Mujahideen in North Africa, they follow its news, and our victory is a victory for them." He was killed while fighting in the western axis of Benghazi in Ganfouda (al-Qa'ida's stronghold) on February 28th, 2016, while battling the LNA.

D

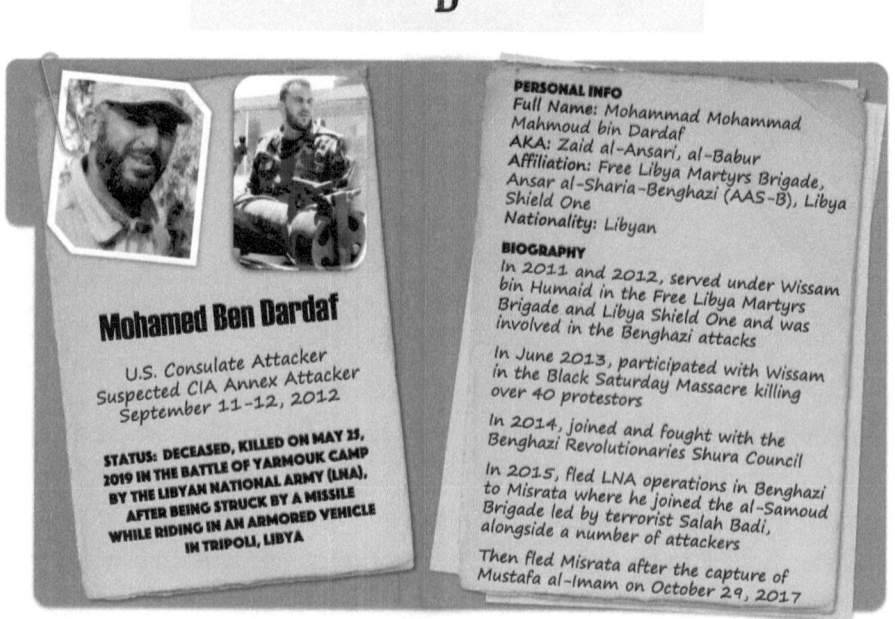

Mohamed Ben Dardaf
U.S. Consulate Attacker
Suspected CIA Annex Attacker
September 11-12, 2012

STATUS: DECEASED, KILLED ON MAY 25, 2019 IN THE BATTLE OF YARMOUK CAMP BY THE LIBYAN NATIONAL ARMY (LNA), AFTER BEING STRUCK BY A MISSILE WHILE RIDING IN AN ARMORED VEHICLE IN TRIPOLI, LIBYA

PERSONAL INFO
Full Name: Mohammad Mohammad Mahmoud bin Dardaf
AKA: Zaid al-Ansari, al-Babur
Affiliation: Free Libya Martyrs Brigade, Ansar al-Sharia-Benghazi (AAS-B), Libya Shield One
Nationality: Libyan

BIOGRAPHY
In 2011 and 2012, served under Wissam bin Humaid in the Free Libya Martyrs Brigade and Libya Shield One and was involved in the Benghazi attacks

In June 2013, participated with Wissam in the Black Saturday Massacre killing over 40 protestors

In 2014, joined and fought with the Benghazi Revolutionaries Shura Council

In 2015, fled LNA operations in Benghazi to Misrata where he joined the al-Samoud Brigade led by terrorist Salah Badi, alongside a number of attackers

Then fled Misrata after the capture of Mustafa al-Imam on October 29, 2017

(47) Mohamed Ben Dardaf, full name Mohammad Mohammad Mahmoud bin Dardaf with aliases Zaid al-Ansari and al-Babur, was from Benghazi. Dardaf's father was also a terrorist and shared the same name, so the biography for the two became convoluted over time. Specifically, the father served in Abu Salim prison, and the son (our attacker) did not.

His uncle, Wissam al-Zaidi, also served in Abu Salim prison and was one of the founders of Ansar al-Sharia-Sirte with several Benghazi attackers.

In 2011, Dardaf joined the Free Libya Martys Brigade, led by Wissam bin Humaid, and later supported Wissam when he led Libya Shield One starting in 2012. On September 11th, 2012, Dardaf was confirmed as an attacker at the U.S. Consulate in Benghazi, and he was also suspected of being an attacker at the CIA Annex in Benghazi on September 12th, 2012 along with Wissam. In the years that followed, Dardaf was involved in several attacks with Wissam, including the Black Saturday massacre on June 8th, 2013, when they killed over 40 protestors outside their Libya Shield One Headquarters in the Abudazira district in Benghazi.

In 2014, he joined the BRSC with Wissam. In the summer of 2014, he fought with several Benghazi attackers in the kick-off to the three-year Battle for Benghazi, pitting terrorists against the LNA. Dardaf escaped LNA targeting operations in Benghazi and fled from Benghazi to Misrata where he had familial connections. In Misrata, along with several other Benghazi attackers, he joined the al-Samoud Brigade led by terrorist Salah Badi. The Brigade was also known as Fakhr or "Pride of Libya" and the Misratan Al Marsa Central Shield Brigade. Badi has been long known to undermine elections in Libya and is best known for destroying the Tripoli International Airport in 2014.

Dardaf then fled Misrata in early November 2017, after several terrorists got spooked after the U.S. capture of Mustafa al-Imam on October 29th, 2017. At that point, he moved on to Ajilat and Sabratha and finally settled in Tripoli. He participated in the battles of Tripoli against the LNA and was killed in the battle of Yarmouk camp on May 25th, 2019, when he was riding in an armored vehicle, struck by a missile on Airport Road in Tripoli. The missile had his name written on it before striking him. Dardaf was publicly mourned and eulogized by several BRSC terrorists from his roots in Benghazi.

(48) Youssef al-Darsi, full name Youssef Ibrahim Mohammed al-Darsi from Benghazi. In 2007, Youssef was detained and sent to Abu Salim prison for supporting global terrorist facilitation networks to include al-Qa'ida. In 2011, during the Libyan revolution, he fought with the Rafallah al-Sahati Brigade. In 2012, he joined AAS-B and served as AAS-B's Commander for the al-Sabri neighborhood. He was known for his brutality, committed terrorist acts, and targeted assassinations against any Libyan in al-Sabri who joined or supported the police, security services, and military. He committed atrocities in Benghazi with terrorists Faraj Azma and fellow Benghazi Consulate attacker Ahmed bin Nasser.

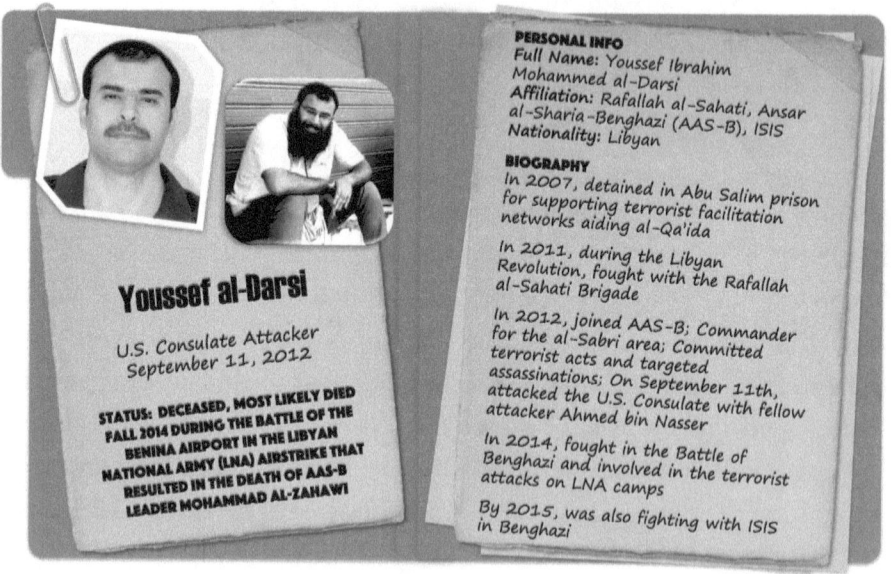

Youssef al-Darsi

U.S. Consulate Attacker
September 11, 2012

STATUS: DECEASED, MOST LIKELY DIED FALL 2014 DURING THE BATTLE OF THE BENINA AIRPORT IN THE LIBYAN NATIONAL ARMY (LNA) AIRSTRIKE THAT RESULTED IN THE DEATH OF AAS-B LEADER MOHAMMAD AL-ZAHAWI

PERSONAL INFO
Full Name: Youssef Ibrahim Mohammed al-Darsi
Affiliation: Rafallah al-Sahati, Ansar al-Sharia-Benghazi (AAS-B), ISIS
Nationality: Libyan

BIOGRAPHY
In 2007, detained in Abu Salim prison for supporting terrorist facilitation networks aiding al-Qa'ida

In 2011, during the Libyan Revolution, fought with the Rafallah al-Sahati Brigade

In 2012, joined AAS-B; Commander for the al-Sabri area; Committed terrorist acts and targeted assassinations; On September 11th, attacked the U.S. Consulate with fellow attacker Ahmed bin Nasser

In 2014, fought in the Battle of Benghazi and involved in the terrorist attacks on LNA camps

By 2015, was also fighting with ISIS in Benghazi

On September 11th, 2012, he attacked the U.S. Consulate in Benghazi and, in its immediate aftermath, was known to brag about his participation in the events within extremist circles. In 2014, he fought with al-Qa'ida-affiliated terrorists in several attempts to overthrow LNA camps. Youssef also linked up with ISIS in Benghazi during this time and fought with them. While he was confirmed to be deceased, there is some discrepancy as to when exactly he was killed. Youssef died in the fall of 2014 during the Battle of the Benina Airport. He most likely was killed in the LNA airstrike that severely injured, AAS-B Leader Zahawi, who succumbed to those injures and died.

(49) **Hamad al-Fakhri,** full name Hamad Noah Younis al-Fakhri from Benghazi. Hamad was a member of the EIJ. In 2007, he was sent to Abu Salim prison for supporting terrorist facilitation networks. Terrorists freed him during the start of the Libyan revolution. In 2011, during the revolution, he joined the Ubaydah bin Jarrah (UBJ) Battalion. Hamad acted as Ahmed Abu Khatallah's driver during the revolution.

In 2012, Hamad joined AAS-B, and on September 11th, 2012, he attacked the U.S. Consulate in Benghazi. On June 15th, 2014, Khatallah was captured in Libya, and in the immediate aftermath, Hamad hid in his father's house, fearing capture. By September 2014, Hamad was comfortable back out in public and joined several terrorist groups in the city to fight the LNA. Hamad was also involved in terrorist efforts to take control of LNA camps in the city.

In February 2017, Hamad was trapped and killed in the 12 Buildings neighborhood in the western axis of Benghazi during the war with LNA. This location was considered the last terrorist stronghold in the western axis after the fall of al-Qa'ida's base in Ganfouda in December 2016. The BRSC used the Saraya Media Center to eulogize Hamad after his death.

Enemies: Brothers in Arms—The Attackers

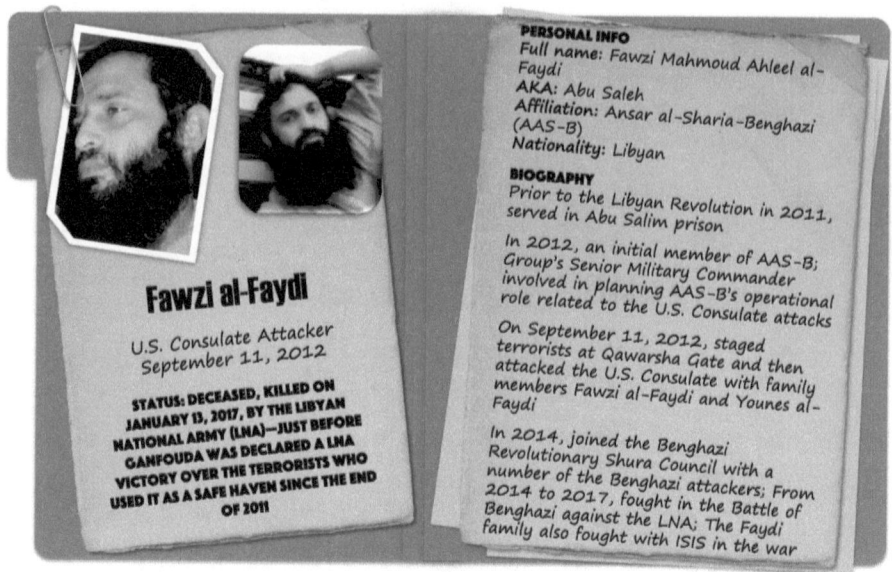

(50) Fawzi al-Faydi, full name Fawzi Mahmoud Ahleel al-Faydi with alias Abu Saleh from Benghazi. Before the Libyan revolution in 2011, Fawzi was a detainee in Abu Salim prison. In 2012, Fawzi was one of the original members of AAS-B and was the group's Senior Military Commander. He was involved in planning AAS-B's operational role in the attack as it related to direct support to al-Qa'ida by providing battle-hardened attackers.

On September 11th, 2012, Fawzi set up the staging at the Qawarsha Gate, where a number of the terrorists met before the attacks, and then all traveled in a convoy to attack the U.S. Consulate in Benghazi. He attacked the Consulate with two family members, Khaled al-Faydi and Younes al-Kish al-Faydi, with alias Abu Muadh (also referred to as "Younes Abu Muadh"). In addition to being an attacker, Younes was a key facilitator for AQIM members who traveled into Benghazi to participate in the attacks.

In 2014, Fawzi joined the BRSC with several Benghazi attackers. He also fought in the Battle of Benghazi with the terrorists against the LNA from 2014 through 2017. On January 13th, 2017, the LNA reportedly killed Fawzi in an airstrike against BRSC in Ganfouda as the al-Qa'ida safe haven had almost completely fallen to the LNA. On January 25th,

2017, the spokesman for the General Command of the LNA announced the liberation of Ganfouda from terrorist organizations.

(51) Khaled al-Faydi, full name Khaled Ali Salem al-Faydi from Benghazi. In 2007, he was arrested and sent to Abu Salim prison for aiding global terrorist networks, including al-Qa'ida. At the time, he was affiliated with the EIJ. In 2011, during the Libyan revolution, he fought with the Qawarsha Brigade. Khaled was instrumental in turning the Qawarsha neighborhood into a terrorist training center, as Gaddafi's former military had abandoned several farms and facilities. Terrorists from all over the world received training in Qawarsha starting in 2011. LNA counterterrorism operations starting in 2014 ended all the terrorist training camps in Benghazi by 2017.

In 2012, Khaled joined AAS-B and led several bombings that targeted police stations and internal security-related facilities in Benghazi. On September 11th, 2012, he staged with a group of terrorists at Qawarsha Gate and then traveled over to attack the U.S. Consulate in Benghazi. Starting in 2014, he fought with terrorists against the LNA in the Battle of Benghazi. In April 2016, he was injured while fighting the LNA in Qawarsha. As LNA made significant gains against the terrorists in Benghazi, Khaled fled to Western Libya.

Khaled's family was full of Benghazi-based terrorists, including Mansour al-Faydi, Viktor al-Faydi, and fellow U.S. Consulate in Benghazi attackers Fawzi al-Faydi and Younes al-Faydi. The family also linked up with ISIS during the 2015 timeframe while fighting the LNA. As of 2022, Khaled was likely in western Libya and traveled between Misrata and Tripoli.

G

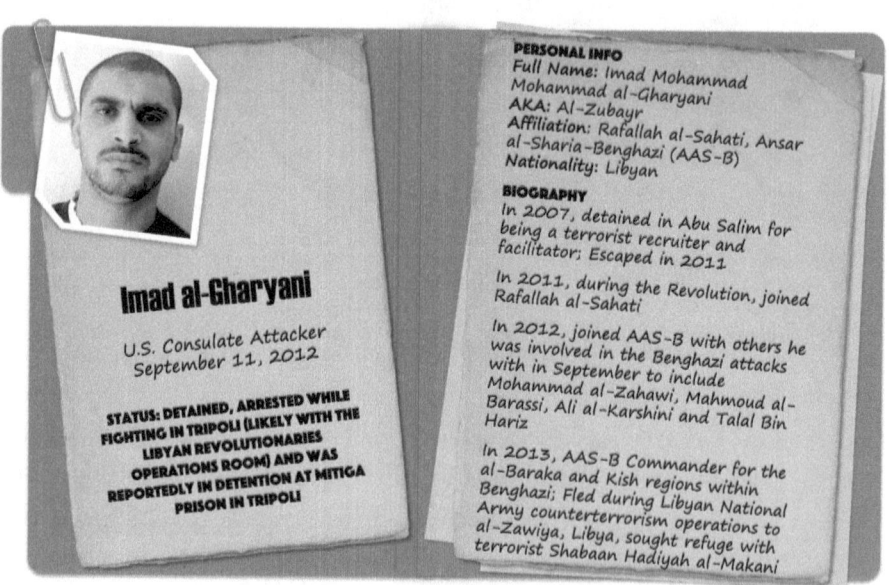

Imad al-Gharyani

U.S. Consulate Attacker
September 11, 2012

STATUS: DETAINED, ARRESTED WHILE FIGHTING IN TRIPOLI (LIKELY WITH THE LIBYAN REVOLUTIONARIES OPERATIONS ROOM) AND WAS REPORTEDLY IN DETENTION AT MITIGA PRISON IN TRIPOLI

PERSONAL INFO
Full Name: Imad Mohammad Mohammad al-Gharyani
AKA: Al-Zubayr
Affiliation: Rafallah al-Sahati, Ansar al-Sharia-Benghazi (AAS-B)
Nationality: Libyan

BIOGRAPHY
In 2007, detained in Abu Salim for being a terrorist recruiter and facilitator; Escaped in 2011

In 2011, during the Revolution, joined Rafallah al-Sahati

In 2012, joined AAS-B with others he was involved in the Benghazi attacks with in September to include Mohammad al-Zahawi, Mahmoud al-Barassi, Ali al-Karshini and Talal Bin Hariz

In 2013, AAS-B Commander for the al-Baraka and Kish regions within Benghazi; Fled during Libyan National Army counterterrorism operations to al-Zawiya, Libya, sought refuge with terrorist Shabaan Hadiyah al-Makani

(52) Imad al-Gharyani, full name Imad Mohammad Mohammad al-Gharyani with alias al-Zubayr was from the al-Baraka area of Benghazi. Imad was arrested for his involvement in terrorism in 2007 after the arrest of terrorist Ali Eshteiwi Dou Posta with alias the Kiwi. Kiwi had reported that Imad was involved in facilitating terrorists to include their recruitment, smuggling, and the forgery of their travel documents. Imad was sentenced to prison for being a terrorist recruiter and facilitating foreign fighters to conflict zones, including Iraq and Lebanon. He escaped Abu Salim prison during the Arab Spring in 2011.

Then during the Libyan revolution, he was a member of the Rafallah al-Sahati Brigade. He then went on to be one of the founding members of AAS-B with Benghazi attack associates Mohammad al-Zahawi, Younes al-

Faydi, Mahmoud al-Barassi, Nasser al-Tarshani, Ali al-Karshini, and Talal Bin Hariz. After his involvement in the September 11th, 2012, U.S Consulate in Benghazi attacks, Imad became the AAS-B Commander for the al-Baraka and Kish regions within Benghazi. After LNA counterterrorism operations commenced in Benghazi, Imad fled to the city of al-Zawiya, where he sought refuge with terrorist Shabaan Hadiyah al-Makani. Imad was arrested fighting in Tripoli, likely with the Libyan Revolutionaries Operations Room, and as of 2022 was reportedly being held in Mitiga prison in Tripoli.

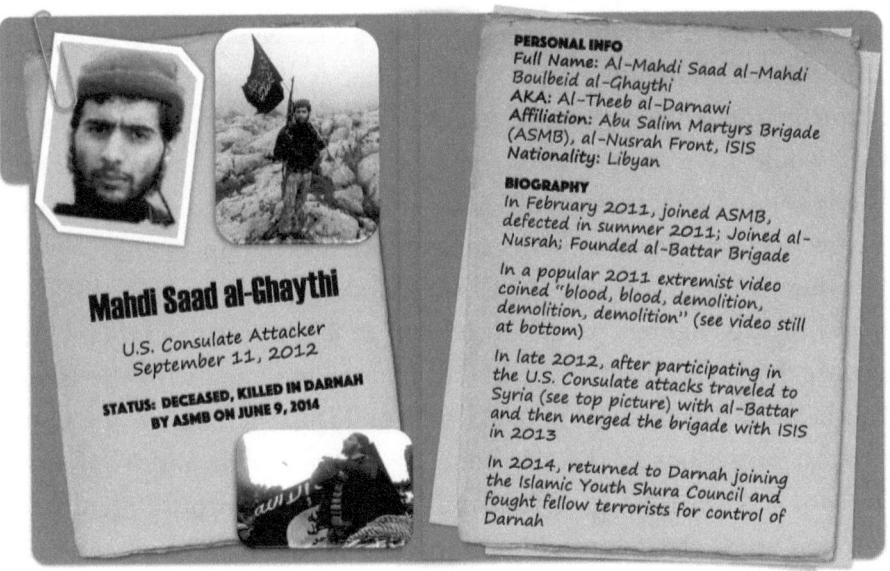

(53) Mahdi Saad al-Ghaythi, full name Al-Mahdi Saad al-Mahdi Boulbeid al-Ghaythi, with alias Al-Theeb al-Darnawi, from the western Shiha neighborhood in Darnah. During the Libyan revolution, he joined the Abu Salim Martyrs Brigade (ASMB) in February 2011. Also, in 2011, Mahdi became jihadi famous for a video he was filmed in threatening blood and death—in the video, he recites the phrase "blood, blood, demolition, demolition, demolition," which caught on and is how the video continues to be referenced.

Being a hardline Islamist, Mahdi defected from ASMB in July 2011 after the group guarded former Gaddafi Official and National Transitional Council's Mustafa Abdul Jalil when he visited Darnah. Fatwas or religious edicts were issued after the incident noting that terrorist affiliates needed to protect Sharia, not politicians trying to depose Sharia. After leaving ASMB, Mahdi supported the al-Nusrah Front. In 2012, he founded the al-Battar Brigade, Syria Branch, with the name variant Katibat al-Battar al-Libi. The Brigade consisted of defectors from ASMB and other extremist groups operating in Benghazi, Ajdabiya, and Tripoli. Mahdi was a key Military Official and member of its Security Committee.

On September 11th, 2012, Mahdi traveled with a group of fighters from Darnah to carry out the attacks on the U.S. Consulate. Several months after the Benghazi attacks, in late 2012, Mahdi traveled to Syria to join other Benghazi attackers and al-Battar Brigade members. These terrorists immediately fled to Syria after the Benghazi attacks fearing the U.S. would have an aggressive response to the death of an Ambassador. While still in Syria, in April 2013, he joined ISIS as the al-Battar Brigade merged its operational activities into the terrorist group.

Then in the spring of 2014, Mahdi, along with many of ISIS's al-Battar Brigade fighters, returned to Libya to join terrorists in their battles against the LNA. In Darnah, Mahdi became affiliated with the Islamic Youth Shura Council (IYSC). As part of IYSC, he carried out several terrorist operations in the city, some against al-Qa'ida and its affiliates. In Darnah at the time, al-Qa'ida and ISIS-affiliated terrorists were at war with one another for control of the city.

On June 9th, 2014, rivalries came to a head when Mahdi was traveling in a vehicle in Aqaba, Shiha al-Sharqiah, near the former Al-Fateh School, opposite the Darnah Wadi. ASMB ambushed his vehicle and assassinated him. On June 10th, 2014, ISIS's al-Battar Brigade issued a statement regarding his death and vowed vengeance for it.

H

(54) Abu Hamza al-Tabawi, real name Ali Mohammad al-Toghi Hammadi with aliases Abu al-Aswad al-Darnawi and Ali al-Darnawi. He was also known as Ali Mohammad Yaqoub al-Tabawi from Darnah. Abu Hamza became involved in terrorism in the early 2000s after joining EIJ. In

2007, he was arrested for supporting global terrorist facilitation networks to include al-Qa'ida, and sent to Abu Salim prison. In 2011, Abu Hamza escaped Abu Salim at the beginning of the Libyan revolution. He returned home and joined the Abu Salim Martyrs Brigade (ASMB) in its Atef Jamal al-Hasadi cell. As part of the operational cell, he was involved in several crimes and acts of terror committed in Benghazi.

Abu Hamza was also associated with AQIM's Mali Group led by fellow Benghazi attackers Mahdi Dango and Jaafar Azzouz. Jaafar's brother Abdel-Qader Azzouz was one of the co-founders of ASMB with Abdel-Hakim al-Hasadi, who fought in Afghanistan with bin Laden and Zawahiri. Jafaar, Abdel-Qader, and a third brother Hamid, likely traveled with Abu Hamza to Benghazi, where they all participated in the attack on the U.S. Consulate on September 11th, 2012.

Abu Hamza was vocal against the LNA and, in 2014, kidnapped two LNA officers who were also members of his tribe, Mohammad al-Tabawi and Ali al-Tabawi. Abu Hamza's younger brother Ibrahim with alias Ito al-Tabawi pledged allegiance to ISIS and was killed in the Battle of the Benina Airport by the LNA in October 2014.

In 2015, the battles between al-Qa'ida and ISIS to control Darnah became more deadly. Abu Hamza remained loyal and aligned with al-Qa'ida. However, after Benghazi attacker Amin Kelfa assassinated ISIS leaders Abdul Nasser al-Aker and Faraj al-Houti on June 9th, 2015, Abu Hamza was worried ISIS would start targeting him next. So, he fled first to Ajdabiya and then to Sirte. And then left Sirte for Ubari, Libya, due to his affiliation with AQIM as the group had a safe haven there, and the fact that he had many cousins residing in the city.

On September 19th, 2019, Abu Hamza was killed in a U.S. airstrike in Murzuq, Libya. The U.S. reported, after the fact, that it had killed eight Libyan ISIS fighters. As such, it seems like it was a failed attempt against ISIS Leader Abdul Qader al-Najdi, who was killed in September 2020 by the LNA. What is clear, though, is that this strike killed AQIM members and not "ISIS fighters" as publicly reported.

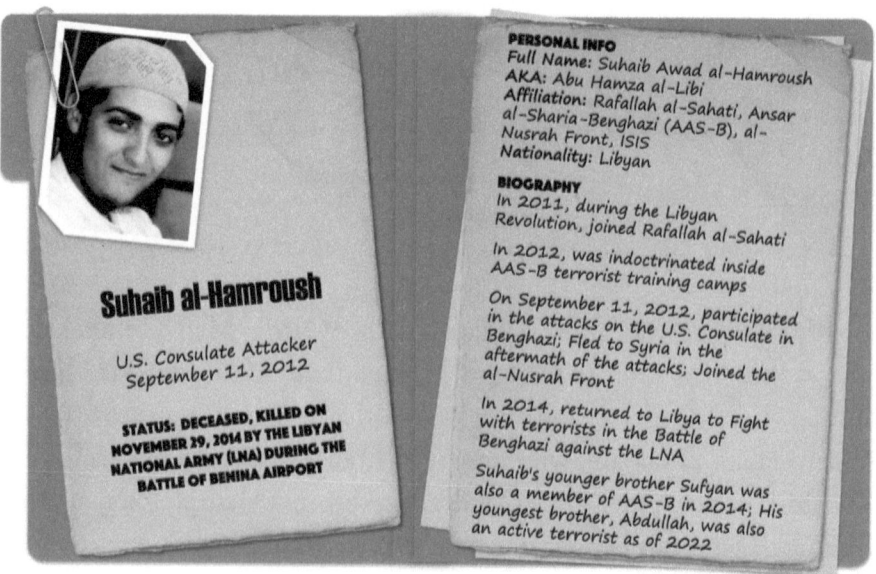

(55) Suhaib al-Hamroush, full name Suhaib Awad al-Hamroush with alias Abu Hamza al-Libi from Benghazi. When the Libyan revolution kicked off, Suhaib was only 21 years old and joined the Rafallah al-Sahati Brigade. In 2012, after the end of the revolution, Suhaib went on to attend terrorist training camps operated by AAS-B and in these camps received indoctrination training, as well. On September 11th, 2012, he participated in the attacks on the U.S. Consulate in Benghazi on behalf of AAS-B. Suhaib fled to Syria in the aftermath of the attacks as many of the attackers feared immediate U.S. reprisals for our Ambassador's killing. In Syria, Suhaib joined the al-Nusrah Front.

In 2014, once the LNA kicked off Operation Dignity, Suhaib returned to fight with the terrorists in the Battle of Benghazi. On November 29th, 2014, he was killed during the Battle of Benina Airport. As an aside, Suhaib's younger brother Sufyan also joined AAS-B. He also has a brother named Abdullah, who, as of 2022, was only 17 years old and was already a terrorist.

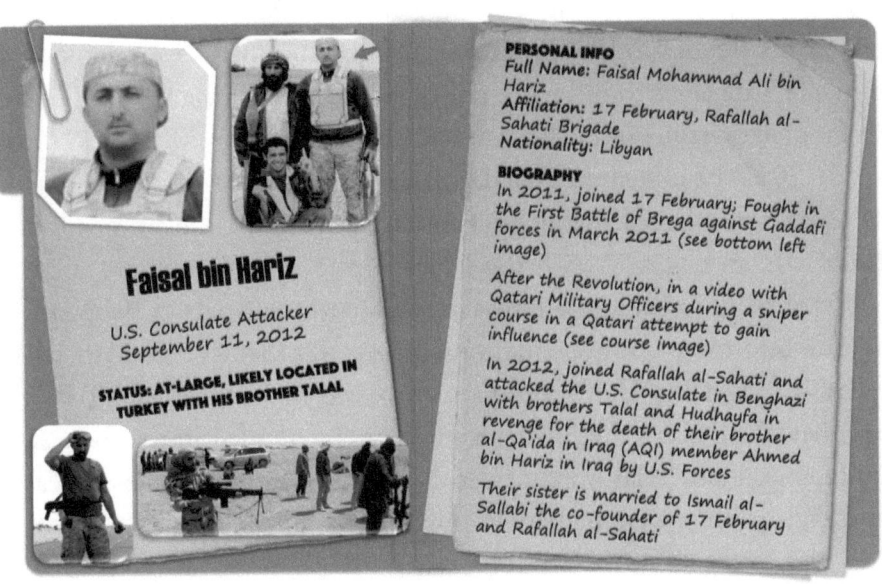

(56) Faisal bin Hariz, full name Faisal Mohammad Ali bin Hariz from Benghazi. Faisal comes from a family of terrorists. His brother Ahmed bin Hariz was an original member of AQI and close to its founder Abu Musab al-Zarqawi. Ahmed was killed while fighting U.S. forces in Iraq in 2004. Two of his brothers participated in the Benghazi attacks with him to avenge Ahmed's death. All the brothers are related to through marriage to 17 February and Rafallah al-Sahati Brigade founder Ismail al-Sallabi, as he was their brother-in-law. At the time of the Benghazi attacks, Ismail was the Military Commander of 17 February and was the second most senior al-Qa'ida operational commander based in the city.

In 2011, during the Libyan revolution, Faisal joined 17 February and fought in the First Battle of Brega against Gaddafi forces in March 2011. Just after the revolution, Faisal appeared in a video with Qatari Military Officers during an operational sniper course with a group of extremists, including Salem al-Mushaiti, Mohammad al-Khafifi, and Mohammad al-Ferjani. The training was filmed by Benghazi attacker Mohammad al-Kawil who was responsible for documenting the activities of al-Qa'ida in Libya. The event hosted high-level leaders from Libyan Battalions who fought the former regime and former Libyan Military officers who defected during the war against Gaddafi. The Qataris had been heavily reliant on the Sallabi family during the revolution, so were looking to find alternate allies to influence policy as it related to the Government of Qatar's relationship with what was becoming a new Government in Libya.

In 2012, he joined the Rafallah al-Sahati Brigade, which again started as an operational cell within 17 February during the Libyan revolution. Faisal was an engineer with the Libyana phone company and used his access at work to geolocate military and security officials in Benghazi for terrorists to carry out assassinations against. On September 11th, 2012, Faisal attacked the U.S. Consulate with two of his brothers, Talal and Hudhayfah. During the incident, Faisal was wounded in the foot. As of 2022, Faisal and Talal were located in Turkey, Hudhayfah was deceased, and their parents were in Tripoli, Libya.

(57) Hudhayfah bin Hariz, full name Hudhayfah Mohammad Ali bin Hariz with alias Abu Bakr al-Shamali was the brother of attackers Talal and Faisal bin Hariz. The bin Hariz brothers had strong connections to AQI and former members of the group as their brother Ahmed bin Hariz was a well-known first-generation AQI member. Ahmad had fought in Iraq and been killed by the U.S. in 2004. During the 2011 revolution, Hudhayfah joined the Rafallah al-Sahati Brigade, and then in 2012, he joined AAS-B. He was the mentee of well-known Benghazi terrorist and fellow attacker Salim Bayou, who was beginning to focus on moving foreign fighters into Syria at the time.

Benghazi: Know Thy Enemy

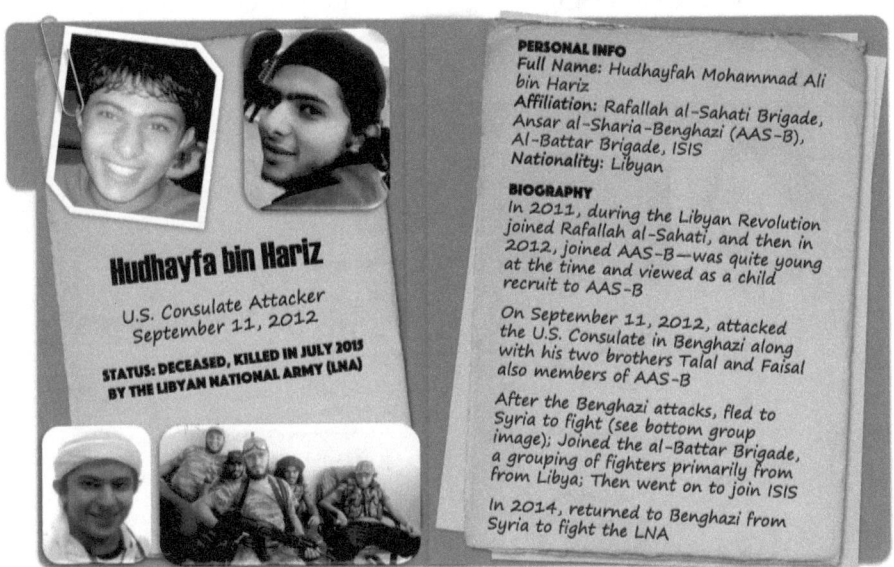

On the night of September 11th, 2012, shortly before the attacks, Hudhayfah departed the Bouzghiba area of Benghazi with three car loads of terrorists possessing light and medium weapons—most of the terrorists were Egyptian Nationals. The Egyptians had reportedly stayed the night of September 10th, 2012, at the home of fellow Benghazi attackers Mahmoud and Yousef al-Awami. Mahmoud had been a Sharia lawmaker in AAS-B, and at the time, he operated between two mosques in Benghazi, the al-Ansar Mosque and the Shuail Mosque. In addition to Mohmoud, Hudhayfah was close to Benghazi attackers Anis al-Houti from Benghazi and Mahdi Saad al-Ghaythi from Darnah.

Immediately after participating in the attacks on the U.S. Consulate, Hudhayfah fled to Syria with several attackers fearing the U.S. would carry out counterterrorism operations in Benghazi to seek justice for our four deceased Americans. While in Syria in 2012, he initially joined the al-Battar Brigade and joined ISIS when the two terrorist groups merged. In approximately 2014, he returned to Benghazi to fight the LNA. He was killed by the LNA in July 2015 while fighting in the Venice Axis in Benghazi. In the aftermath of his death, Hudhayfah was eulogized by the Al-Tanasuh channel, which broadcasted from Turkey and was affiliated with the Dar Al-Iftaa in Libya, led by Al-Sadiq al-Gharyani, the Grand Mufti of Libya.

(58) Talal bin Hariz, full name Talal Mohammad Ali bin Hariz from Benghazi, was living in Sweden before the revolution. It is unclear how he resided in Sweden as he was a former EIJ member. In the mid-2000s, he was detained and sent to Abu Salim prison for supporting global terrorist facilitation networks to include for al-Qa'ida. In 2007, he was given a release order and released from prison. Several Benghazi attackers were released at this time, as there were large protests in Benghazi objecting to the detentions at Abu Salim prison. Gaddafi's son Saif al-Gaddafi secured the release of Talal and others by lobbying through his Gaddafi International Foundation for Charity Associations to quell the protests.

In 2011, when Talal returned from Sweden to Libya to fight during the revolution, he joined 17 February, where his brother Faisal was also a member. Their sister was also married to Ismail al-Sallabi, who, as noted, was one of the founders of 17 February. In 2012, Talal joined AAS-B. At the time of the attacks, he had been a Sharia Judge for AAS-B in the Qawarsha neighborhood in southwestern Benghazi, primarily occupied by terrorists. He worked closely with AAS-B members Mansour al-Faydi and Benghazi attacker Fawzi al-Faydi at the time.

On September 11th, 2012, Talal and his two brothers, Faisal and Hudhayfa, participated in the Benghazi attacks. Talal was visible on video footage (later played on the Alan channel) while outside the U.S. Consulate at the time of the attacks. The LNA killed Hudhayfa during counterterrorism operations in Benghazi. Another brother, Ahmad bin Hariz had been one of the original AQI members and was a close associate of Abu Musab al-Zarqawi. U.S. forces killed Ahmad in Iraq in the 2003/2004 timeframe.

In 2014, Talal was a senior leader in the BRSC, founded by many Benghazi attackers. On July 29th, 2014, he fought with al-Qa'ida-affiliated terrorists when they overran and captured the Libya Special Forces (Saiqa) Camp in Benghazi. Other Benghazi attackers were involved in the incident, including Wissam bin Humaid, Mohammad al-Zahawi, and Jalal al-Makzoum. Talal also became associated with ISIS over the next year and fought with the terrorist group in Benghazi, as well. In 2017, after al-Qa'ida was defeated in Benghazi by LNA, Hariz relocated to Turkey. As of 2022, his brother Faisal was also reported to live there.

Enemies: Brothers in Arms—The Attackers

(59) Ali Ouni al-Harzi, full name Ali bin al-Tahar bin al-Falih al-'Awni al-Harzi with alias Abu Zubayr al-Tunisi was from the Ariana region in Tunisia. Ali and his brother Tariq al-Harzi, full name Tariq bin al-Tahar bin al-Falih al-'Awni al-Harzi with alias Abu Umar al-Tunisi were key foreign fighters and facilitators linked to both al-Qa'ida and ISIS networks across North Africa, Iraq, and the Levant.

Historically, both brothers were AQI members, joining soon after Abu Musab al-Zarqawi formed the group in 2004. In 2005, Tunisian authorities arrested Ali for planning terrorist attacks on behalf of AQI, and he was sentenced to just 30 months in prison. Due to Ali's detention, Tariq moved up faster in the ranks and served as a trusted facilitator for al-Qa'ida. Several years later, Tariq was acclaimed as one of the first foreign fighters to join ISIS starting in 2007.

In Tunisia, the brothers were close associates of Seifallah Ben Hassine, a long-time associate of bin Laden. Hassine founded Ansar al-Sharia-Tunisia (AAS-T) in 2011 during the Arab Spring. While Ali was aligned with Hassine as he served in a role trafficking terrorists to Syria and weapons to Tunisia, his primary affiliation nearing the time of the 2012 Benghazi attacks was with the Al-Ghuraba Brigade. This Brigade was an AQIM Brigade under the Uthman bin Affan Brigade, founded by fellow Benghazi attackers Anis al-Houti and Mahmoud al-Wahishi.

In 2011, Ali was operating an Al-Ghuraba Brigade training camp on the Libyan side of the Tunisian-Libyan border, where he would train Tunisian members and a mix of other foreign fighters for AQIM. In 2012, when AAS-B was formed, AQIM allowed them to take over these camps as they were planning joint terrorist attacks together, including the U.S. Consulate in Benghazi attack and the In Amenas, Algeria attack. Ali also became a member of AAS-B and continued training terrorists in the camp for al-Qa'ida and AAS-B. Not much changed regarding the camp's function when it moved to AAS-B control, with Ali still sending most of his trainees to fight in Syria and Iraq.

Ali had been an attacker at the U.S. Consulate in Benghazi on September 11th, 2012. Tariq was involved in the September 14th, 2012, attack against the U.S. Embassy and an American school in Tunis,

Tunisia. After the Benghazi attacks, fearing American reprisals, Ali fled to Turkey. He was captured in Turkey on his way to Syria, with a travel associate, due to both possessing fake identification. Most of the Benghazi attackers that fled in the attack's aftermath had real Libyan passports but with false names and some false nationalities. AAS-B had connections within the Government of Libya which provided real Libyan passports to not just Libyan al-Qa'ida terrorists but to a host of other nationalities who participated in the attacks, including some German terrorists. The passports were prepared before the attacks to allow the attackers to flee Libya unabated in quick succession.

In December 2012, Turkey deported Ali back to Tunis, where the FBI interrogated him. On January 8th, 2013, he was released due to a "lack of evidence," even as the FBI had ample evidence. The Muslim Brotherhood in Tunisia, through the political party Ennahda, was heavily involved in securing Ali's release. The U.S. Government likely did not put in an extradition request for Ali. As a show of defiance, Hassine's AAS-T posted a video statement lauding Ali's freedom.

On January 23rd, 2013, Secretary Clinton testified before the Senate Foreign Relations Committee, noting that the Tunisians had "assured" the United States that Ali was "under the monitoring of the court." Just 14 days later, on February 6th, 2013, Ali, with fellow Benghazi attacker Boubaker al-Hakim assassinated a Tunisian opposition party leader Chokri Belaid. The next day, on February 7th, 2013, our then newly appointed CIA Director John Brennan incorrectly reported to Congress that the U.S. Government "didn't have anything on" Ali; therefore, his release was not problematic. Not only did we have the intelligence of Ali's involvement, but he even posted that he attacked the Consulate on his social media.

On July 25th, 2013, Ali and Boubaker assassinated a second opposition leader in Tunisia, Mohamed Brahmi. Ali was not tracked after his release and became a leader of ISIS in Syria. In July 2015, then-Senate Judiciary Committee Leader Senator Chuck Grassley wrote a letter to James Comey, then FBI Director, and Loretta Lynch, U.S. Attorney General, querying whether the FBI mishandled Ali's case. We can respond to that request for you, YES.

By 2013, brother Tariq was assisting in financing ISIS in its key warzones of Iraq and Syria and reportedly bringing millions into the group. By this time, Tariq had also become the ISIS Leader for its suicide-bombing network, where he ran a pipeline of potential suicide bombers between Syria and Iraq. Tariq also was directly involved in External Operations on behalf of ISIS and successfully recruited a host of Europeans and North Africans to attack the West.

After supporting AAS-T's assassinations in Tunisia, Ali fled the country and became directly involved in supporting ISIS, alongside his brother, primarily in Iraq. He was responsible for recruiting foreign fighters to Syria and was involved in ISIS's hostage program. On April 14th, 2015, the State Department designated Ali a terrorist; however, the designation excluded that he was involved in the U.S. Consulate attacks. A common issue for years was that our government refused to report terrorists involved in the Benghazi attacks. On June 15th, 2015, a U.S. airstrike killed Ali in Mosul, Iraq, and on the next day, June 16th, 2015, a U.S. drone strike killed Tariq in Shaddadi, northeastern Syria.

Faraj Shakku
U.S. Consulate Attacker
September 11, 2012
STATUS: DECEASED, KILLED ON JUNE 13, 2018 IN A U.S. AIRSTRIKE IN BANI WALID, LIBYA

PERSONAL INFO
Full Name: Faraj Musa al-Hasnawi
AKA: Shakku
Affiliation: 17 February, Ansar al-Sharia-Benghazi (AAS-B), ISIS, al-Qa'ida
Nationality: Libyan

BIOGRAPHY
During the Gaddafi Regime, served in the Libyan Army in its General Administration support division; In 2011, joined 17 February

In 2012, was a AAS-B Commander and worked for the Blue Mountain Group (BMG) which provided security at the U.S. Consulate compound

In 2014, pledged allegiance to ISIS where he led the Bouhdima axis; Libyan National Army (LNA) lost 700 military officers in Bouhdima from 2014-2017

In 2016, founded the Benghazi Defense Brigades (BDB); In May 2017, carried out the Brak al-Shati massacre against the LNA, over 140 killed

(60) **Faraj Shakku**, real name Faraj Musa al-Hasnawi from the Bouhdima neighborhood in Benghazi, and he had a Tunisian mother. He was referred to as Shakku, a name his father had been known by. Shakku was

affiliated with al-Qa'ida. Of note, Shakku was one of the few terrorists involved in the Benghazi attacks that had served prior in his country's military. He had served in the Libyan Army under Gaddafi in its General Administration support division.

In 2011, during the Libyan revolution, Shakku was a member of 17 February under Ismail al-Sallabi. In 2012, he was a contractor working with the Blue Mountain Group (BMG), which provided the unarmed guards at the U.S. Consulate in Benghazi. On September 11th, 2012, he attacked the Consulate. From 2011 to 2014, Shakku was also a suspect in many al-Qa'ida-led assassinations in the city of Benghazi.

In 2014, Shakku joined the BRSC with many of the Benghazi attackers. Also, in 2014, he pledged allegiance to ISIS in Benghazi, where he led the Bouhdima axis in battles against the LNA. The LNA tragically lost 700 military officers in Bouhdima alone during the Second Libyan Civil War. From 2014 to 2017, the LNA lost 8,000 officers total just in the eastern cities of Benghazi and Darnah. Shakku also appeared in a video released by the Bushra Media Agency, affiliated with ISIS.

Shakku was also a founder of the Benghazi Defense Brigades (BDB) and fought in support of the BDB in the al-Sabri and Ganfouda neighborhoods in southern Benghazi. He participated in a number of the attacks on oil ports and oil facilities and was wanted for his involvement in the BDB Brak al-Shati massacre in May 2017, where over 140 people were killed.

As LNA began to take control of Benghazi, Shakku fled briefly to southern Libya. He likely was temporarily under the protection of AQIM. Like other attackers, he then moved on to Western Libya, where he decided to join the al-Wefaq militia based near Misrata. He then moved on to support fighting in western Libya, including key cities, Tripoli, Sorman, and Sabratha, against the LNA, reportedly in support of and sanctioned by the Government of Libya.

He participated in battles against the LNA in Tripoli, under the leadership of terrorist In June 2018, al-Qa'ida announced the deaths of Shakku and close associate and fellow Benghazi Consulate attacker Ahmed al-Tajouri after they were killed on June 13th, 2018 in a U.S. airstrike near Bani Walid, Libya. The strike was only reported as targeting

AQIM killing "one terrorist"—it was never reported that two Benghazi Consulate attackers were killed.

(61) Ahmed Hazaa, full name Ahmed el-Sayed Hazaa Hassan with alias Osama from Egypt. In 2006, Ahmed was arrested in Tripoli after traveling from Egypt with an Ansar Allah cell. The cell intended to travel to Iraq to fight U.S. forces and to join AQI. When the cell first arrived in Tripoli, they linked up with Benghazi attacker Malik al-Khazmi who was operating a separate cell for Ansar Allah in Tripoli. Malik was arrested as well. In 2011, all five cell members escaped from Abu Salim prison and fought in the Libyan revolution. In 2012, Ahmed joined AAS-B. On September 11th, 2012, Ahmed attacked the U.S. Consulate in Benghazi with the four members of his cell from 2006.

Separately in 2012, after the revolution, a group of Egyptians were arrested in Sirte, Libya. The detainees reported that Ahmed and his cell members were affiliated with AAS-B. It was also reported that they were trafficking weapons to Mali on behalf of AAS-B and that they were also key members of AQIM's Mali Group. As of 2022, the status of Ahmed was unknown, but he was suspected of being at-large in Libya.

(62) **Hani al-Hawari** with aliases Bin Laden and Mahrouqa from Darnah, was a member of the Abu Salim Martyrs Brigade during the Libyan revolution in 2011. Before the revolution, Hani had been a Libyan Government security official in Darnah and then made friends with some reformed persons from his investigations. These relationships led him to first become a criminal and then a terrorist. As terrorist organizations took over sections of key cities throughout Libya in the aftermath of the revolution, many Libyan men joined terrorist groups when they likely would not have under different circumstances.

In 2012, he participated in the U.S. Consulate attacks with a grouping of Darnah-based terrorists and, at the time, was a member of AAS-B. In the immediate aftermath of the attacks, he stayed in Benghazi. After the attacks in 2012, Hani was on video with Tunisian Benghazi attacker Hossam bin Hassouna al-Zaytouni, breaking into the Benghazi Council of the Benghazi Security Directorate where the Lightning Battalion was based.

Hani then traveled to Syria and joined the al-Battar Brigade with several fellow Benghazi attackers. He joined ISIS while in Syria, as well. In 2013, Hani returned from Syria to Libya via Turkey. He was arrested at the Benina International Airport in Benghazi upon arrival for his involvement in the Security Directorate attack. It is unclear how long he was held, but it was not long.

While in Darnah in April 2014, he was riding in the vehicle of AAS-B member Mohamed Suleiman Tajouri when he was assassinated. Hani then fought in the Second Libyan War and was again on video in July 2014, when al-Qa'ida terrorists took control of the LNA-affiliated Special Forces Saiqa camp. His last battle was the Battle of the Benina Airport, which lasted from August 2014 until October 2014, when the LNA killed him.

(63) Anis al-Houti, full name Anis Faraj Abd al-Salam Mohammad al-Amin al-Houti with alias Abu al-Muhajir from Benghazi. Anis was a member of AQIM at the time of the attacks. In 2003, after the U.S. invasion of Iraq kicked off, Anis departed Libya to fight in Iraq. He first traveled to Algeria and stayed there to participate in a terrorist training camp. He never made it to Iraq at that time as intended. In Algeria, he joined the Salafist Group for Preaching and Combat (GSPC) and then became a member of AQIM.

In 2011, after the Libyan revolution kicked off, Anis returned to Libya with fellow Benghazi attacker, Mahmoud al-Wahishi, and the two formed the Othman Battalion. Unlike many Libyan battalions, which were almost solely made up of Libyan nationals, Othman had terrorists from many nationalities across the Sahel and reportedly from Europe. After the Benghazi attacks, Anis continued his terrorist activities in Libya, leading to confrontations with the LNA during its counterterrorism operations starting in Benghazi in May 2014. According to his September 2014 eulogy, Anis was killed by LNA during the Battle of Benina Airport.

(64) **Mustafa al-Imam**, full name Mustafa Mohamed al-Imam from Zliten, Libya, was an ethnic Palestinian. Reportedly, Mustafa was from a family of extremists in Palestine, with their activities dating back to the 1990s. He was captured in 2007 for links to terrorist organizations in Benghazi and sent to Abu Salim prison. In 2011, during the Libyan revolution, Mustafa fought with 17 February and then moved to Benghazi for the first time. He lived in the al-Laythi neighborhood near Ahmed Abu Khatallah, and it was assessed he and Khatallah became connected while serving in Abu Salim together.

In 2012, Mustafa joined AAS-B and was reportedly a preacher or legal theorist with the group. He had likely done some membership recruiting for AAS-B, as well, traveling between Darnah, Ajdabiya, and Sirte. On September 11th, 2012, Mustafa and Khatallah showed up to the attacks late after Khatallah received calls from terrorist allies that an attack was occurring at the U.S. Consulate in Benghazi. After arriving at the Consulate, neither Mustafa nor Khatallah entered the Consulate compound until after all the actual al-Qa'ida terrorists fled. When they did enter, they looted some items not of any real value, as the Consulate could not maintain any classified holdings on the premises.

Starting in 2014, Mustafa participated in the Battle of Benghazi against the LNA. There are a few images online of Mustafa in this period where he appears to be working as the security detail for the camera crew of Channel Al-Nabaa. In the pictures, he is in front of the 17 February camp where the BRSC, established in 2014, was based at the time. While numerous Benghazi Consulate attackers founded BRSC, Mustafa is seen protecting a press team, not on the other side with BRSC.

In December 2016, Mustafa reportedly was involved in the Oil Crescent attacks carried out by the Benghazi Defense Brigades. Mustafa was injured fighting in Benghazi in 2017 when the LNA made significant progress against the terrorists and their Benghazi safe haven. He fled Benghazi aboard a sea dredger to Misrata and then traveled to Turkey for medical treatment. After receiving treatment in Turkey, he returned to Misrata.

Mustafa was captured soon after by U.S. forces and the FBI on October 29th, 2017. To be blunt, his capture was a bit perplexing as it seemed to be used to prop up the fake claim that Ahmed Abu Khatallah was the "Mastermind", as there were actual Consulate attackers in his almost immediate vicinity in Misrata when his capture occurred, including Mohammad al-Manfi and Mohamed Ben Dardaf. Dardaf not only attacked the U.S. Consulate, but was also suspected of attacking the CIA Annex. In addition, by 2017, several senior attackers were already detained in Libya, and not even one was extradited to the United States. It is unclear why a rendition of Mustafa was necessary with more high-value targets easily accessible. His capture even confused terrorists, as did the capture of Khatallah, who was called the "scapegoat" in Libya for being falsely charged as the Mastermind.

On January 23rd, 2020, Mustafa was sentenced to 19 years for looting the Consulate, however, he did have crimes against Libyans and the LNA that would warrant a lengthy sentence and even the death penalty. As an aside, when searching for persons from Libya linked to al-Qa'ida, the only identified al-Qa'ida or al-Qa'ida-affiliated terrorist with a name similar to Mustafa's in Libya was al-Qa'ida member Abdul-Monsef Mohammad Hassan al-Imam, who holds British citizenship. There was no information indicating the two were related.

(65) Mohammad Jaafar, full name Mohammad Hassan Rajab Jaafar from Benghazi. Jaafar was a former Abu Salim prisoner. In 2011, during the Libyan revolution, he joined 17 February and fought with the group in Benghazi. In 2012, he joined AAS-B, where he was involved in attacks on police stations and cemeteries across several neighborhoods in Benghazi. On September 11th, 2012, he attacked the U.S. Consulate in Benghazi with his brother, Osama Jaafar. No updated information was found on either terrorist after they participated in the Consulate attacks, and as of 2022, the status of both Mohammad and Osama was unknown.

Enemies: Brothers in Arms—The Attackers

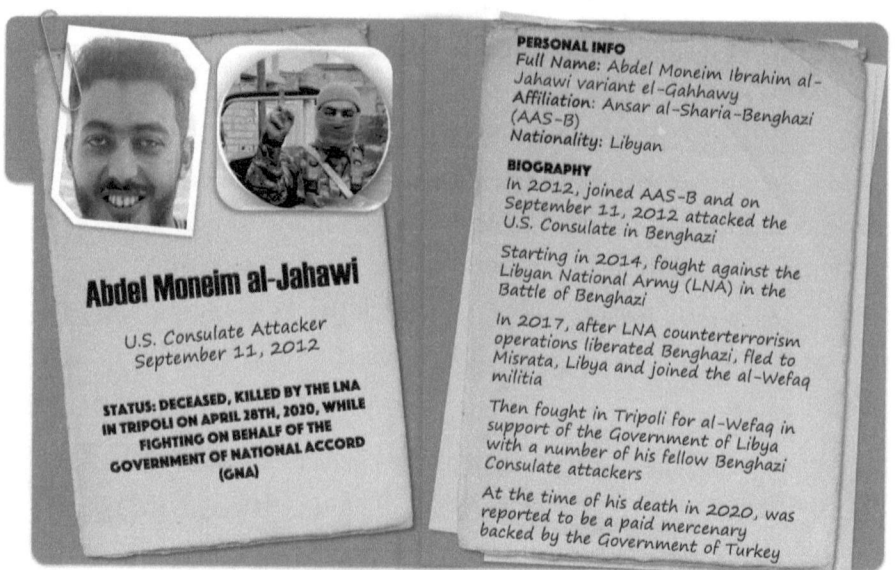

(66) Abdel Moneim al-Jahawi, full name Abdel Moneim Ibrahim al-Jahawi variant el-Gahhawy, from Benghazi. Jahawi's past prior to the Libyan revolution was unknown, but he was only in his early 20s at the time, so he likely had no previous jihadist history. In 2012, he joined AAS-B, and then, on September 11th, 2012, he attacked the U.S. Consulate in Benghazi. After the attacks, he stayed in Benghazi and fought against the LNA in the Battle of Benghazi. As Haftar's counterterrorism operation started making real gains throughout 2016, Jahawi in 2017 decided to flee to Misrata, Libya, where he joined the al-Wefaq militia. This militia was also focused on battling the LNA but in western Libya.

The LNA killed Jahawi in Tripoli on April 28th, 2020, while he fought on behalf of the Government of National Accord (GNA). Reportedly, he was also a paid mercenary fighting on behalf of contracts levied between Turkey and Libya at the time for security assistance.

K

(67) Munther al-Kubti, full name Munther Mansour al-Kubti from the al-Rahba neighborhood in Benghazi. In 2011, during the Libyan revolution, he fought with 17 February. In 2012, he joined Libya Shield One under CIA Annex attack Mastermind Wissam bin Humaid. In the group, Munther was a senior official and the security official in charge of the group's gates into its headquarters in the Ganfouda area. This is important as Ganfouda was also al-Qa'ida's base at the time, and al-Qa'ida was a key partner to Wissam in Ganfouda.

Further, it was reported that some of the al-Qa'ida operatives that participated in the U.S. Consulate in Benghazi attacks, like attacker Khaled al-Saqzli, had sought refuge in the Libya Shield One headquarters the night of September 11th, 2012, after completing the U.S. Consulate attacks. As such, on September 11th Munther was suspected of providing support for attackers at the U.S. Consulate attacks. He was also suspected of directly participating in the attacks on the CIA Annex in Benghazi on September 12th, 2012.

Starting in 2014, Munther fought with the terrorists against the LNA during the Second Libyan Civil War. That same year, he joined the BRSC and pledged allegiance to ISIS. While a member of ISIS, he was involved in a high-profile attack against the Zawiya Martyrs Brigade that same year. On June 2nd, 2014, terrorists attacked the Tabalino Base in Qar

Yunis, Benghazi. This led LNA and the Saiqa Special Forces to respond to assist the Zawiya Martyrs Brigade. The attack Munther helped lead resulted in at least 23 people being killed and 97 injured.

Munther ended up leaning away from ISIS and aligning back with al-Qa'ida-affiliated terrorists. He then fled Benghazi to Misrata as LNA was making counterterrorism gains against the terrorists in Benghazi. In Misrata, he joined the 1st Company of the Eagle Brigade. This led him to travel to Tripoli to battle the LNA, and on April 4th, 2019, he joined the al-Wefaq militia.

As a quick sidebar, many Benghazi attackers later joined the al-Wefaq militia, which supported the Libyan Government in battling the LNA, mainly in Tripoli. Of note, one of the terrorists who attacked our U.S. Embassy in Tunisia in the aftermath of the Benghazi attacks on September 14th, 2012, also joined this brigade. That was Abu Yahya Zakaria with the real name Bilal bin Yusuf bin Mohammad al-Shawashi, a Tunisian who was arrested after the 2012 incident and later released. He traveled to Syria, first joining the al-Nusrah Front and then joining ISIS. It was in 2019 when he traveled to Libya to fight for al-Wefaq. While this book does not get into the Tunisia issues, it is important to note how many of these attackers crossover and link up in North Africa. First due to historic relationships held by Seifallah Ben Hassine from Ansar al-Sharia-Tunisia (AAS-T), then from fighting with AQI and AQIM, and lastly with relationships formed through ISIS.

On March 30th, 2020, Munther was killed by the LNA while fighting in the Ramla axis in Libya's capital of Tripoli. The BRSC announced his death and eulogized him.

(68) Atef al-Karami, full name Atef Mustafa Muhammad al-Karami from Benghazi. Atef was introduced to terrorism through his brother, Aha Abdullah al-Karami. Aha was a member of the Martyrs Group and carried out a suicide attack in Benghazi in 1998. Atef also has a younger brother who joined ISIS and became a suicide attacker in Sirte. Atef first joined Hamza Bouchertila's group, and when Hamza became a leader in the Martyrs Group, Atef also joined the group. In 2005, with fellow

Martyrs Group member and U.S. Consulate attacker Muftah al-Ammari, the two attempted to rob the Trade Bank and were arrested. He was sent to Abu Salim prison.

Atef al-Karami

U.S. Consulate Attacker
September 11, 2012

STATUS: LIKELY AT-LARGE, WAS ARRESTED IN AUGUST 2016 IN SIRTE, LIBYA AND REPORTEDLY MANAGED TO ESCAPE AND FLEE TO WESTERN LIBYA AT THE TIME

PERSONAL INFO
Full Name: Atef Mustafa Mohammad al-Karami
Affiliation: Egyptian Islamic Jihad (EIJ), Martyrs Group, Ansar al-Sharia-Sirte (AAS-S), ISIS
Nationality: Libyan

BIOGRAPHY
In 1998, arrested for being a member of EIJ, sent to Abu Salim prison

In 2001, released and went on to join the Martyrs Group; In 2005, was arrested with fellow Consulate attacker Muftah al-Ammari while robbing a bank, sent back to Abu Salim

In 2011, during the Revolution, fought in Sirte; Committed a mass atrocity killing 38 inmates at a local prison

In 2012, participated in the Benghazi Consulate attack with the Rafallah al-Sahati Brigade

In June 2013, helped establish the AAS-S, and then joined ISIS becoming the group's media official

In 2011, during the revolution, Atef first traveled to Sirte as his family had moved there after his arrest in 2005. Atef was known to have committed mass atrocities during the revolution in the Battle of Sirte. On October 20th, 2011, he was involved in killing 38 prisoners, with Human Rights Watch identifying him as a perpetrator of the incident. In Sirte, he was committing attacks with terrorists Ahmed Ali al-Tir, Abdelhadi Zarqun, Fayez Attiyah, Fawzi al-Ayyat, Wissam al-Zaidi (uncle of Benghazi attacker Mohamad ben Dardaf), and Benghazi attacker Abdul-Ati Abu Sitta. It would be with these same terrorists that Atef co-founded Ansar al-Sharia-Sirte (AAS-S) approximately two years later, starting in 2013. At the revolution's end, Atef temporarily returned to Benghazi, where he joined the Rafallah al-Sahati Brigade.

In June 2012, Atef participated in AAS-B's first annual sharia conference in Benghazi. Then on September 11th, 2012, he participated in the attacks on the U.S. Consulate in Benghazi. Approximately a year after the attack, Atef relocated back to Sirte, where he led terrorist activities for AAS-S.

As ISIS moved into Sirte, Atef joined early, becoming the group's media official. He was also a righthand to Bahraini terrorist Turki al-Binali.

In August 2016, during the battles to liberate Sirte from the terrorists, a Misrata militia captured Atef and Fawzi al-Ayyat, who by then were two of the most wanted ISIS terrorists in Sirte. A few days later, Fawzi appeared in an investigation video without Atef, who reportedly managed to escape and went into hiding in western Libya. As of 2022, Atef was at-large, and his location was unknown.

Hassan al-Karami
U.S. Consulate Attacker
September 11, 2012

STATUS: DECEASED, ON AUGUST 11, 2016, KILLED IN THE GIZA BAHARIYA DISTRICT OF SIRTE, LIBYA DURING OPERATION AL-BUNYAN AL-MARSOUS

PERSONAL INFO
Full Name: Hassan Mohammad Abdullah al-Karami
AKA: Abu Muawiya, Abu Qatada al-Ansari
Affiliation: Al-Qa'ida, Ansar al-Sharia-Benghazi (AAS-B), ISIS
Nationality: Libyan

BIOGRAPHY
In late 2011, supported al-Qa'ida efforts to establish a base in Benghazi and participated in terrorists acts in Benghazi on behalf of the group

In 2012, was one of the founding members of AAS-B in its Religious Committee; On September 11th participated in the attack on the U.S. Consulate in Benghazi; Fled in 2013

By late 2014, pledged allegiance to ISIS in Sirte becoming the senior Mufti

In August 2015, delivered the order to execute a dozen Ferjani tribesmen whose headless bodies were then left on public display in Sirte

(69) Hassan al-Karami, full name Hassan Mohammad Abdullah al-Karami with alias Abu Muawiya from Benghazi. Hassan had a younger brother named Abu Bakr, whose alias was Muawiya, who died fighting with ISIS in Sirte. Hassan's alias was in honor of him. There was reporting that Benghazi attackers Hassan and aforementioned Atef al-Karami were brothers, however, there is no evidence to support this claim.

In late 2011, as al-Qa'ida started to cement a base in Benghazi after the Libyan revolution, Hassan committed several terrorist acts in Benghazi for the group. In 2012, Hassan supported the creation of the umbrella organization AAS-B and initially was a member of the group's Religious Committee. He was present at the group's kickoff event, its first annual Sharia Conference, in June 2012. On September 11th, 2012, he participated in the attack on the U.S. Consulate in Benghazi.

In 2013, Hassan and his family fled Benghazi for Sirte, where he pledged allegiance to ISIS in Sirte in 2014. In Sirte, he was the head of a Diwan al-Dawah wal Masajid wal-Awqaf, the Administration of Dawa Activity, Mosques, and Religious Endowments. He was also affiliated with the Ribat al-Imami Mosque in Sirte, and renamed it the "Abu Musab al-Zarqawi Mosque" in honor of the founder of ISIS and AQI.

Before the formal establishment of ISIS in Sirte, Hassan had formed a solid relationship with Bahraini terrorist Turki al-Binali, whom locals referred to as the "godfather" and the true founder of ISIS in Libya. In the phase before ISIS was established in Libya, Turki also had a close relationship with al-Qa'ida in the Islamic Maghreb's (AQIM) Mali Group. At least a dozen terrorists from the Mali Group participated in the U.S. Consulate in Benghazi attacks, including Libyan Mali Group members Mahdi Dango, Mansour al-Shalaali, Jaafar Azzouz, Abu Hamza al-Tabawi, Khaled al-Ammari, Taher al-Awami, Abdul Hamid al-Shaeri, Hassan al-Shukri, and Hashem Bousidra.

Hassan terrorized the residents of Sirte in many ways. For one, he produced ISIS publications that summoned former security officials in the city. Essentially, Sirte residents were expected to turn these men in to be "tried" and then executed by ISIS. In August 2015, Hassan participated in the massacre in the eastern Sirte district known as "neighborhood three." In this act of terrorism, Hassan gave the order for ISIS terrorists to behead 12 Ferjani tribal members, whom ISIS hung on crosses on display in the city. Hassan and crew then executed 22 other residents of Sirte while they lay wounded in the hospital and set fire to the hospital.

In March 2016, Hassan was visible on news coverage present alongside patrons in the front rows of a ceremony set up by al-Qa'ida in Tripoli's Martyrs' Square. The ceremony was a prayer vigil to mourn the loss of al-Qa'ida fighters in the eastern cities of Libya, primarily Benghazi and Darnah. While this book focused solely on the U.S. Consulate in Benghazi attacks, which used eastern Libyan-based al-Qa'ida networks, it is important to note that al-Qa'ida also had (and continues, as of 2022, to have) a foothold in western Libya, including in the capital of Tripoli and within Libya's Government. So be aware that a whole other cast of al-Qa'ida characters in the West exist and is not reflected herein.

On August 11th, 2016, Hassan was killed in the Giza Bahariya district of Sirte during Operation al-Bunyan al-Marsous. This operation was launched by Libya's Government of the National Accord (GNA) against ISIS in Sirte.

(70) Ahmed bin Nasser Karim, with aliases Abu Abdullah and Khadrawat from the al-Sabri neighborhood in Benghazi. In 2011, during the Libyan revolution, Ahmed was a member of the 17 February. In 2012, Ahmed joined AAS-B. In AAS-B, Ahmed was closely associated with AAS-B leader Mohammad al-Zahawi, killed in 2014; Mahmoud al-Barassi, killed in 2019; Mansour al-Faydi, with alias Abu al-Layth, who died in the attack on the Sidra area oil ports on December 7th, 2016; and Khaled al-Aqouri with alias Abu Othman who was at-large in southern Libya and had a severed foot.

On September 11th, 2012, Ahmed attacked the U.S. Consulate in Benghazi with fellow AAS-B member Mahmoud al-Barassi. In 2014, Ahmed and a group of terrorists, including Mansour al-Faydi, defected from AAS-B. They pledged allegiance to ISIS when Mahmoud al-Barassi was named the ISIS Leader of Benghazi. Throughout 2014, Ahmed fought with the BRSC, as well.

In ISIS, Ahmed was referred to as the "Butcher of ISIS" as he was known to behead and dismember bodies during his terrorist acts and assassinations. In 2014, Ahmed beheaded and mutilated the corpse of LNA soldier Saleh Awad al-Warad during the initial phase of the Battle of Benghazi. In 2015, he dismembered a resident with alzheimer's, Bashir Mustafa al-Agha, who ISIS charged with being in contact with the LNA. He also killed and mutilated the corpse of Omar al-Ati, for being associated with the LNA battles in the al-Sabri axis. In 2017, after the LNA dealt heavy losses to ISIS in al-Sabri, Ahmed fled with Mahmoud al-Barassi to the desert in southern Libya.

In September 2017, Ahmed appeared in an ISIS publication lauding the August 2017 ISIS attack on an LNA checkpoint in al-Jufrah, Libya. In the attacks, Ahmed and fellow terrorists killed nine LNA soldiers. The LNA soldiers were not killed while fighting. ISIS took them hostage, with some executed at close range, others beheaded, and the last had their throats slit. In the ISIS issue, Ahmed is carrying one of the decapitated heads belonging to LNA Major Ali al-Ghadban, the Commander of the 131st Battalion. Several terrorists closely affiliated with Ahmed and the Benghazi U.S. Consulate attack network appeared in the same ISIS issue, including Mahmoud al-Barassi, Khaled al-Aqouri, Mahdi Dango, and Al-Zubayr al-Tunisi.

On February 3rd, 2018, Ahmed was killed near Zella, Libya, during fighting between ISIS and the Al-Jafra Operation Room of the Libyan Government of the National Accord (GNA). Ahmed was killed with fellow ISIS member Khalifa Fadhil al-Mansouri al-Obeidi, with the alias "The Chechen." Khalifa was a prior member of the DMSC before choosing ISIS over al-Qa'ida when terrorists had to pick a side in the war between the two groups in Darnah.

(71) Ali al-Karshini, full name Ali Mohammad Mohammad Ibrahim al-Karshini with alias Abu Jabal from Benghazi. When Ali got involved in terrorism in the early 2000s, it was with the Libyan branch of the EIJ. He was arrested in 2007 for his involvement in supporting global terrorist networks, to include al-Qa'ida. He was detained in Abu Salim prison. Most of these terrorists were only out of jail because they escaped at the start of the Libyan revolution.

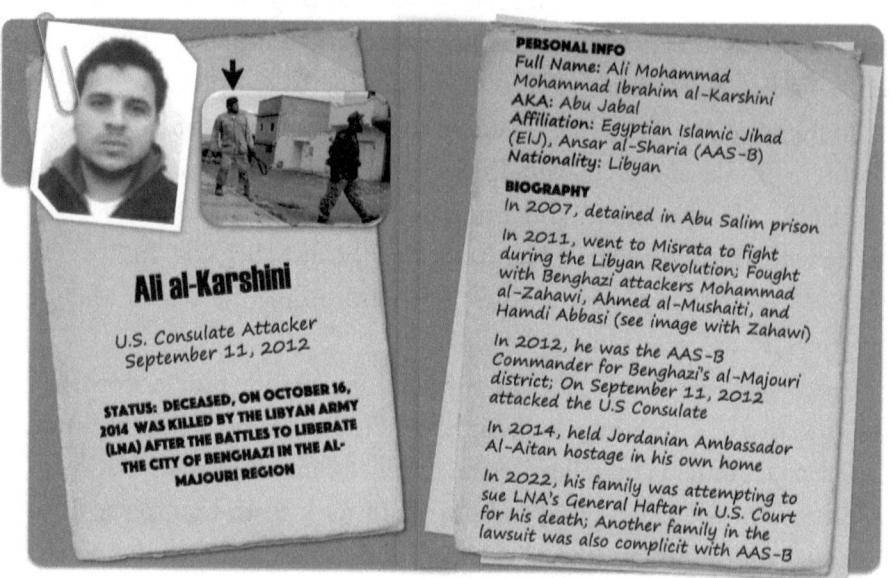

Ali al-Karshini
U.S. Consulate Attacker
September 11, 2012

STATUS: DECEASED, ON OCTOBER 16, 2014 WAS KILLED BY THE LIBYAN ARMY (LNA) AFTER THE BATTLES TO LIBERATE THE CITY OF BENGHAZI IN THE AL-MAJOURI REGION

PERSONAL INFO
Full Name: Ali Mohammad Mohammad Ibrahim al-Karshini
AKA: Abu Jabal
Affiliation: Egyptian Islamic Jihad (EIJ), Ansar al-Sharia (AAS-B)
Nationality: Libyan

BIOGRAPHY
In 2007, detained in Abu Salim prison

In 2011, went to Misrata to fight during the Libyan Revolution; Fought with Benghazi attackers Mohammad al-Zahawi, Ahmed al-Mushaiti, and Hamdi Abbasi (see image with Zahawi)

In 2012, he was the AAS-B Commander for Benghazi's al-Majouri district; On September 11, 2012 attacked the U.S Consulate

In 2014, held Jordanian Ambassador Al-Aitan hostage in his own home

In 2022, his family was attempting to sue LNA's General Haftar in U.S. Court for his death; Another family in the lawsuit was also complicit with AAS-B

In 2011, Ali went to Misrata to fight against Gaddafi's forces during the Libyan revolution. He fought alongside Benghazi attackers Mohammad al-Zahawi, Ahmed al-Mushaiti, Hamdi Abbasi, and Mohammad al-Manfi. He also fought with terrorist Mohammad al-Ferjani, In 2012, Ali assisted in founding AAS-B and was the AAS-B Commander for Benghazi's central al-Majouri district. He led assassinations against many military and security officials, and civilians. Ali also led attacks on police stations and the Benghazi Security Directorate.

On September 11th, 2012, he attacked the U.S. Consulate in Benghazi. He attacked the Consulate with several terrorists he fought with in Misrata, including Zahawi, Manfi, and Abbasi. Then after Jordanian Ambassador Fawaz al-Aitan was kidnapped on April 15th, 2014, in Tripoli and transferred to Benghazi, Ali held the Ambassador hostage at his family home. As noted previously, the Ambassador was released in exchange for an AQI prisoner on May 13th, 2014.

Starting in 2014, Ali fought with the terrorists against the LNA in the Battle of Benghazi. As the LNA was moving towards the al-Majouri area to begin counterterrorism operations to liberate the neighborhood from terrorists, Ali led efforts to stop LNA at the neighborhood's main entry point. LNA killed Ali on October 16th, 2014. During an LNA raid on his family home, millions of Libyan Dinars and U.S. dollars were confiscated, as well as a cache of weapons and munitions and then documents affiliated with the business dealings of AAS-B.

Ali's family members, from this same property where the Ambassador was held hostage, have sued LNA's General Khalifa Haftar in the U.S. court system for the killing of Ali. This case sets a dangerous precedent when we allow our judiciary system to be misused by the families of terrorists. Of note, there is an additional family in this same case suing Haftar, the Suwaid Family. Unlike the Karshini family, who had only one terrorist family member (AAS-B member Ali), the Suwaids had three AAS-B terrorists in the immediate family and another famous al-Qa'ida terrorist in their extended family. The U.S. Department of Justice designated AAS-B a terrorist organization on January 13th, 2014.

In terms of the Suwaid family, the father of this family, Abdul Salam bin Suwaid, and his two sons, Mustafa Abdul Salam bin Suwaid and Khaled Abdul Salam bin Suwaid were members of AAS-B. The two sons were also members of the Rafallah al-Sahati Brigade during the revolution. The father and his sons lead the assassination cell for AAS-B in the Bouhdima neighborhood. When we were in Benghazi, this family was so extreme that they flew the black flag belonging to AAS-B at their family home.

The family was carrying out high-profile assassinations in a neighborhood with many supporters of the army and security apparatuses. As LNA's plan for Operation Dignity was gearing up, most of the area's residents pledged their support to the LNA and issued a public statement not to fight against the LNA in Bouhdima. The LNA kicked off Operation Dignity in 2014, and when they reached Bouhdima, the residents witnessed gunfire coming out of the Suwaid family home directed at the LNA. The residents became incensed, including members of the "Guardians of Blood," an organization set up to seek justice for innocent family members who were assassinated since 2011 by terrorists in Benghazi.

The residents formed a mob outside the family home and surrounded it. Gun battles ensued between the residents and the Suwaid family. The residents then allowed a temporary ceasefire for the women and children inside to evacuate safely. These family members were then banned from Bouhdima. The AAS-B terrorist father Abd al-Salam, two AAS-B terrorist sons, Khaled and Mustafa, and an additional son named Ibrahim were killed.

The LNA did not kill the Suwaids, but Haftar was still being sued for their deaths as of 2022. As the Suwaid terrorists had also joined the al-Qa'ida-affiliated BRSC when it was formed in 2014, BRSC issued a statement. Essentially, BRSC threatened the public and noted it would punish all persons involved in attacking the Suwaid family home. An English translation follows: "The time for retribution from you has come, and we will have another matter with you. You will soon know the consequences of what you did to our brother Abd al-Salam and his sons."

For those long in the counterterrorism community, you are wondering why a Libyan named "Suwaid" was familiar; that is because the Suwaid's cousin is Salem Saad bin Suwaid (full name Salem Saad Salem bin Suwaid with alias Abu al-Habib). Salem was involved in the October 28th, 2002, assassination of U.S. diplomat Larry Foley. Foley was assassinated in Amman, Jordan, while working with the United States Agency for International Development (USAID). The mastermind of this attack was Abu Musab al-Zarqawi.

As a brief background, in 1990, when Salem was living in Peshawar, Pakistan, he lived in the home of Abu Yahya al-Libi. He married the cousin of Abdullah Azzam in Denmark. For those who followed historic terrorist facilitation networks through the Pakistan and Afghanistan border region, Salem was the head of al-Qa'ida's Miram Shah guesthouse in the city of Khowst, Afghanistan, and also the head of the Abu al-Ezz guesthouse. His brother Abdullah was a senior member of EIJ. His cousin, Awad Mohammad al-Zawari, was involved in the Libyan Branch of EIJ with Benghazi attacker Ali al-Karshini. The Jordanians executed Salem on April 6th, 2004.

Benghazi: Know Thy Enemy

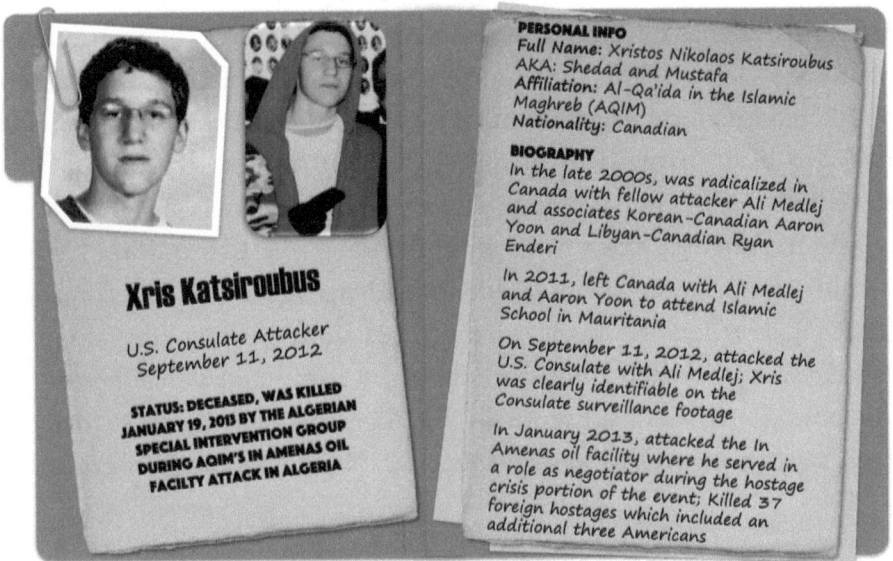

(72) Xris Katsiroubus, full name Xristos Nikolaos Katsiroubus with aliases Shedad and Mustafa from London, Ontario, Canada. Xris was a member of AQIM and fought in the al-Mulathameen Battalion. The battalion was led by U.S. Consulate in Benghazi attacks Mastermind MBM.

Xris, along with fellow Benghazi attacker Ali Medlej, and two additional Canadian associates, Korean-Canadian Aaron Yoon and Libyan-Canadian Ryan Enderi with alias Mujahid were radicalized in Canada. Enderi was the first to leave Canada for Libya in 2010. Medlej, Xris, and Aaron left Canada to attend Islamic School in Mauritania in 2011. Aaron was captured in Nouakchott, Mauritania, on terrorism charges and later deported back to Canada, where he served less than two years. Separately, we have no reporting of the involvement of Enderi in North African terrorist attacks. According to him, after relocating to Tripoli, he lost contact with Medlej, Xris, and Aaron.

On September 11th, 2012, led by AQIM operational commander Abu Bara al-Jazairi, Xris and Ali Medlej attacked the U.S. Consulate in Benghazi. Xris was captured on the Consulate's surveillance footage. Then starting on January 16th, 2013, Abu Bara, Xris, Ali, and another AQIM operational planner Abdul Rahman al-Nigeri, carried out the three-day

177

attack and hostage crisis at the Tigantourine gas facility (also referred to as the In Amenas oil facility) near In Amenas, Algeria. The attack killed 37 hostages from 9 different countries, including 3 Americans. The attack also killed one local Algerian guard based at the facility. Abdul Rahman had also been involved in planning the Benghazi attacks but did not travel to Benghazi to participate directly in the attacks.

At In Amenas, Xris was in an operational role that quickly became a public communications role due to his language ability as a Canadian national. He handled many of the discussions between the al-Qa'ida attackers, and outside company officials, security and police personnel, and hostage negotiators. Eventually, the Algerians came in and carried out a raid against the AQIM hostage takers, and both Xris and Ali were killed.

(73) Abu Yaqin al-Libi, real name Mohammad Fawzi al-Senussi al-Kawafi from Benghazi. Abu Yaqin was just 20 years old when he participated in the attack on the U.S. Consulate. At the time, he served as a bodyguard to the Leader of AAS-B, Zahawi. In the aftermath of the attacks, on September 19th, 2012, the BBC interviewed Zahawi, and Abu Yaqin is visible in the video performing protective duties for the leader. Before the attacks, he was also filmed threatening suicide attacks while riding in a sports utility vehicle (SUV) in Benghazi.

After the attacks, he joined the al-Battar Brigade, which essentially became the Libyan wing of ISIS in Iraq and Syria. While primarily Libyans, it did host some allied terrorists, including members of Ansar al-Sharia-Tunisia (AAS-T). Members affiliated with this Brigade were later involved in supporting the November 13th and 14th, 2015 coordinated terrorist attacks in Paris, France—killing 130; and the May 22nd, 2015 attack at the Ariana Grande concert in Manchester, United Kingdom—killing 22. Abu Yaqin died on April 15th, 2014, before both events, under suspicious circumstances. After his death, al-Battar's media wing eulogized him for being an ISIS martyr.

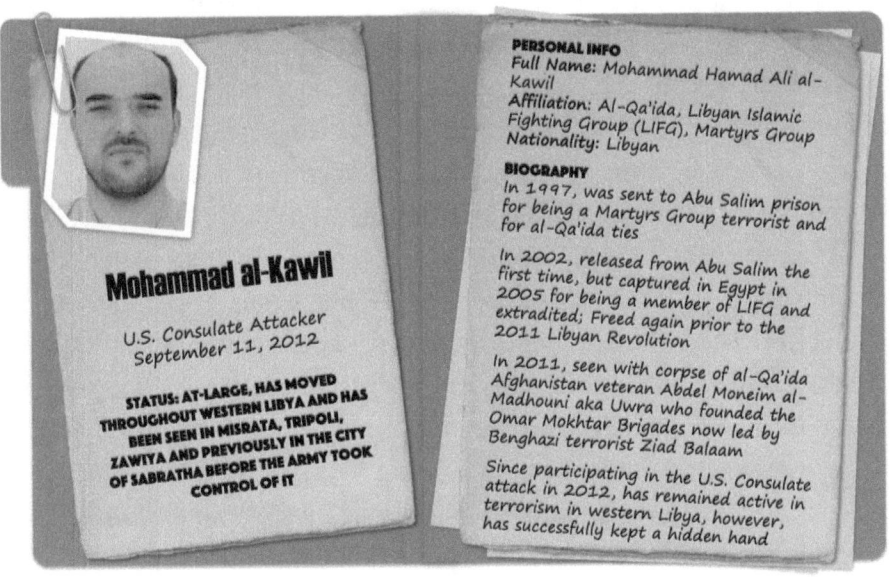

(74) **Mohammad al-Kawil,** full name Mohammad Hamad Ali al-Kawil from the Sidi Hussein area of Benghazi. He was affiliated with al-Qa'ida, previously the LIFG, and currently likely leads the Martyrs Group. This organization was set up in the aftermath of the Abu Salim Prison massacre on June 29th, 1996, where 1,270 prisoners were believed to be killed. The Martyrs Group decided to follow the teachings of Osama bin Laden and al-Qa'ida's ideology and have been affiliated with al-Qa'ida for decades. The Martyrs Group was based in Benghazi, but also operated a European cell.

Kawil links back to the Sudan days when Bin Laden was seeking refuge in the country as he was connected to Bin Laden's Al-Aqiq Company. While there is a lot of focus on relationships forged in Afghanistan, the relationships established in Sudan were just as enduring. Bin Laden and his network established contact with several Benghazi attackers during his Sudan era, including Kawil, AAS-B Leader Zahawi, Ziad Balaam, Sufyan bin Qumo, and Faraj al-Chalabi.

Kawil was sent to serve in Abu Salim prison in 1997 for his association with the Martyrs Group and connection to fellow members Adel al-Awami and Khaled al-Darsi. Inside prison, he formed a close relationship with LIFG members Ahmed Qanoun and Salem al-Mabrouk al-Obaidi, who talked him into joining LIFG. Kawil was released from Abu Salim for the first time in 2002. However, he was not free long as Egyptian authorities captured him in 2005 for being a member of LIFG and deported him back to Libya, where he likely re-entered Abu Salim's prison population in 2006. He was not seen again until his release during the Arab Spring in early 2011.

Then in 2011, he was seen in video footage in front of the corpse of a former al-Qa'ida Senior Leader for Afghanistan, Abdel Moneim al-Madhouni. Madhouni went to fight the Soviets in Afghanistan in the 1980s. In 2011, during the Libyan revolution, Madhouni founded the Omar Mokhtar Brigade. After he died in 2011, it was taken over by terrorist Hussein al-Fitouri who was killed in 2016. As of 2022, Benghazi attacker, Ziad Balaam now leads it. Kawil and Ziad were in Sudan together and then maintained a close friendship while in Abu Salim Prison. They had worked together to instigate riots in prison after Kawil's first release in 2002. As of 2022, both Kawil and Ziad were at-large.

Circling back to historical networks, Kawil and Ziad were a part of a communications network with leaders in al-Qa'ida to ensure continuity of communications. The network had been led by Bashir al-Fiqi with alias Mohammad Ismail, and Abu Abd al-Rahman al-Faki. Faki, a previous resident of Britain, was the al-Qa'ida Leader for Misrata. His relationships with the network linked back to Abdul Rasoul Sayyaf's group in Afghanistan and the GIA in Algeria. Others who took advantage of this network throughout the years included: Misbah Abdul Hafeez al-Ammari, Mustafa Ahmed ben Dardaf, Salem Mohammad al-Aqouri, Ibrahim Boujldin, Ibrahim Ghaith al-Aqouri, Ahmed Saad al-Aqili, and Nizar Saleh al-Senussi. Several of these terrorists were killed fighting U.S. forces in Iraq on behalf of AQI, including Misbah Abdul Hafeez al-Ammari, Salem Mohammad al-Aqouri, and Ibrahim Boujldin.

(75) Amin Kelfa, full name Amin Ali Meloud Kelfa from Shiha in eastern Darnah. In 2011, during the Libya revolution, Amin joined the most prominent militia in Darnah at the time, the Abu Salim Martyrs Brigade (ASMB). Amin had a long jihadist history where he has been involved in terrorism, murder, assassinations, kidnappings, and arms trafficking.

In 2008, Amin assassinated the Head of the Anti-Terrorism Department at Libya's Internal Security Service (ISS), Colonel Naji Fadhil al-Awami. ISS was the Libyan organization most comparable to the FBI. Amin assassinated the Colonel with terrorist Abdullah Habil in the vicinity of Azzouz Island near Darnah. In 2011, he also carried out another assassination in Darnah of a Libyan nurse named Hamdi Juma'a al-Shalawi, a high-profile event in the city.

On September 11th, 2012, Amin traveled with several Darnah-based terrorists to participate in the attacks on the U.S. Consulate in Benghazi.

He had a history of capture and release, either "escaping" or finding persons within the Libyan Government to advocate for his release. In 2012, he was captured with terrorist Murad Magdy al-Sabaa. Both were released, and Murad went on to carry out a high-profile assassination in 2015 of Darnah resident Abdel-Wahhab al-Darsi who defended a local family attacked by ISIS. In what seems to be a separate event, Amin was captured again in 2012 in Sousse, Libya, and carried out an armed escape that freed him.

In 2013, Amin started the Islamic Army in Darnah with Youssef bin Taher, the uncle of Benghazi attacker Ali bin Taher. Youssef bin Taher's title was Chief of Staff of this "Army." Benghazi attacker Ramadan Trabelsi was also a senior leader in the Islamic Army. The "Army" was treated like a quasi-governmental organization by the Libyan Government under Libyan Prime Minister Ali Zeidan, whereas the group was allocated millions of dollars to support operations. The operations were terrorist operations.

In December 2014, Amin flip-flopped and fought against the Islamic Army with ISIS when ISIS attempted to take over the town of Ain Mara, Libya. He fought in the axes of Al-Dahr al-Hamr near the city of Darnah. The fighting force aligned with the Islamic Army was al-Qa'ida-affiliated and fell under the Undersecretary of the Ministry of Defense in the Zeidan government. However, the Libyan Government had been funding these al-Qa'ida-affiliated militia elements since 2012, during the Government of Abdurrahim el-Keib using Deputy Defense Minister al-Siddiq al-Mabrouk al-Ghaythi, who decided which Islamist groups received funding. Al-Ghaythi would be the individual to help kick off funding for the Benghazi Defense Brigades (BDB), which Amin would later join in 2016. Al-Ghaythi's son Abdul-Malik al-Siddiq al-Ghaythi was a member of al-Qa'ida and died fighting in Al-Dahr al-Hamr.

By 2015, Amin realized his heart was with al-Qa'ida joining the group and its local arm, the DMSC. He then began to battle ISIS. As al-Qa'ida's main focus in eastern Libya since 2011 were assassination campaigns, that's what Amin focused on as the battles in Darnah between al-Qa'ida and ISIS escalated. He led the assassinations of many senior ISIS members, including Abdul Nasser al-Aker and Faraj al-Houti, in

June 2015. Then he murdered former terrorist colleague Murad (from his 2012 detention), as Murad had killed Amin's terrorist brother Abdul-Muhaymin Ali Kelfa during the Battles of Darnah in December 2015. Not to be one-upped, ISIS then targeted Amin's other brother Tamim Ali Kelfa with an improvised explosive device (IED) which led to his death in February 2016.

After the death of a second brother, Amin knew ISIS was closing in on him, and in June 2016, he fled Darnah onboard a marine dredger to the Qasr Hamad Port in the city of Misrata. He then traveled south to the al-Jufrah area. He linked up with some of the old U.S. Consulate in Benghazi attacks crew, including 17 February and Rafallah al-Sahati Brigade founder Ismail al-Sallabi and Benghazi attackers Ahmed al-Tajouri and Faraj Shakku. These terrorists fought at the time under the banner of the BDB, as mentioned prior.

As a member of BDB, Amin took part in many terrorist incidents orchestrated by the group, the first being the July 2016 attack against the cities of Ajdabiya, al-Maqron, and al-Jelida. On September 10th, 2016, he was arrested again in the town of Harawa, Libya, with 18 terrorists from Darnah. Amin ended up being released due to support from a Leader in the BDB, Colonel Mustafa Suleiman al-Sharkasi. Sharkasi was an Islamist, well-known in the city of Benghazi, and connected to many of the key senior leaders involved in the U.S. Consulate attacks. By December 2016, Amin was back in action and participated in a port facility attack with BDB.

On May 18th, 2017, he and fellow BDB terrorists carried out the Brak al-Shati massacre when they attacked the Brak al-Shati Airbase operated by the LNA. At the airbase, 141 people were killed, including 103 LNA soldiers. On June 14th, 2018, Amin, with BDB terrorists and allied Chadian militia fighters, attacked an oil facility setting it ablaze outside Ras Lanuf, Libya. This incident led to the closure, at the time, of two critical oil terminals in As-Sider and Ras Lanuf. In September 2018, he participated in deadly clashes in southern Tripoli.

As central Libya, where al-Jufrah was located, was no longer hospitable for terrorists, Amin relocated farther south in Libya. He supported al-

Qa'ida and AQIM allies along Libya's southern route near the cities of Sabha and Ubari. On January 27th, 2019, Amin was captured again while co-located with Benghazi attacker Adel Ahmad al-Abdali who was ambushed by the LNA in Sabha. Adel, at the time, was a Field Commander for AQIM and died during the ambush. Amin was again released. In early July 2022, Amin was arrested again by a security authority in the Majer area, near the city of Zliten, Libya. He will likely not remain in custody, and should be extradited immediately.

(76) Marei Zoghbi, full name Marei Abdel-Fattah Khalil from Benghazi. Zoghbi had several aliases over the years, including Zoghbiyeh, Larzg Ben Ila, Lazrag Faraj, F'raji Di Singapore, F'raji Il Libico, Mohammad el-Besir, Mohamed Lebachir, Farag, and Fredj.

In 2001, Zoghbi traveled to Afghanistan, where he first joined the LIFG at its Kabul camp, and then he joined al-Qa'ida. In 2002, Zoghbi linked up with Tunisian Imed ben Mekki Zarkaoui, who was also linked to Seifallah Ben Hassine, who fought in Afghanistan and who supported the Tunisian angle related to the U.S. Consulate attacks in Benghazi. This angle included supplying and facilitating terrorists for the attacks and then assisting attackers as they fled in the aftermath of the attacks.

Through Zarkaoui, Zoghbi ended up in Milan, Italy. In 2007, Zoghbi was arrested in Italy for planning terrorist attacks with al-Qa'ida's Milan Cell and Ansar al-Islam. The members of his cell at the time included Al-Azhar ben Khalifa ben Ahmed Rouine, Faraj Faraj Hussein al-Sa'idi, Riadh ben Belkassem ben Mohamed al-Jelassi, Radi Abd el-Samie Abou el-Yazid el-Ayashi, Faycal Boughanemi and Nessim ben Romdhane Sahraoui. The cell was linked to several groups supporting al-Qa'ida recruitment efforts in Europe, North Africa, Asia, and the Middle East. With Ansar al-Islam, Zoghbi was recruiting Europeans to travel to Afghanistan and Iraq to train and fight with al-Qa'ida. On December 20th, 2007, Zoghbi was given a 6-year sentence which was overturned on November 20th, 2008.

In 2011, during the Libyan revolution, Zoghbi returned to Libya from Europe and joined the Rafallah al-Sahati Brigade in Benghazi. Zoghbi fought in Afghanistan with the Brigade's Leaders Mohammad al-Gharabi and Ismail al-Sallabi. Besides leading Rafallah al-Sahati the night of the attacks, they were the two most senior al-Qa'ida Commanders based in the city of Benghazi at the time. On September 11th, 2012, Zoghbi attacked the U.S. Consulate in Benghazi with his historic al-Qa'ida associates.

In 2014, Zoghbi joined the al-Qa'ida-affiliated BRSC and fought with the terrorists against the LNA in the Battle of Benghazi. He fled LNA counterterrorism operations in Benghazi for Darnah, where he was under the protection of the al-Qa'ida-affiliated DMSC, which was BRSC's primary ally in Darnah. He co-founded the al-Qa'ida-affiliated al-Mourabitoun organization in Darnah with U.S. Consulate attack Mastermind MBM and Egyptian terrorist Hesham Ashmawy. On October 8th, 2018, the LNA carried out a raid on the leadership of al-Mourabitoun, capturing Zoghbi, Heshawy, and Bahaa Ali Ali Abu al-Maati.

(77) Malik al-Khazmi, full name Al-Sadiq Salem al-Khazmi with aliases Abu Ibrahim and al-Hajj Ibrahim, was from Bani Walid, Libya. Khazmi was a former member of AQI. Over the years, he also had relationships with Ansar Allah, the LIFG, Ansar al-Sharia-Sirte (AAS-S), and ISIS.

Enemies: Brothers in Arms—The Attackers

In the mid-2000s, Khazmi traveled to Iraq to join AQI and fight U.S. forces. In 2004, he returned to North Africa after a couple of years of fighting and fled to Egypt as Libyan authorities were trying to capture him for being an AQI terrorist. In Egypt, Khazmi linked up with the terrorist group Ansar Allah which had been around since the original Afghan Mujahideen days.

As a quick background, on August 8th, 1996, Ansar Allah was formed in Libya. The group consisted of the Libyans who had fought in Afghanistan under several different Afghan Mujahideen leaders. In the late 1980s, approximately 1000 Libyans traveled to fight in Afghanistan.

One influential leader was Gulbuddin Hekmatyar, the Leader of Hezb-e Islami Gulbuddin (HIG). Ansar Allah members Mohammad Omar Hassan Bayou and Yusuf Mohammad al-Qamati with alias Issa Abdul Qayyum fought with him during the Soviet-Afghan War. In late 2016, it was Bayou's son, Jaafar Mohammad Bayou, who died in Mastermind Wissam Bin Humaid's vehicle when the terrorist was targeted. Wissam later died as a result of this attack. As of 2022, Qamati was living in Benghazi. Another influential leader related to Ansar Allah was Abdul Rasoul Sayyaf, who led Ittehad-al-Islami (also referred to as the Islamic Union) in the vicinity of Jalalabad Province in Afghanistan. In Jalalabad, a number of what would become future members of Ansar Allah had also trained or fought with Osama bin Laden and other future members of al-Qa'ida in this locale.

For Ansar Allah, Khazmi also trained Egyptian terrorists in weapons and explosives tradecraft. In 2006, Khazmi left Egypt to return to Libya, where he trained terrorists in Libya before being facilitated to Iraq to fight. In November 2006, he was captured and charged with terrorism for leading an operational cell in Tripoli for Ansar Allah. He was sent to Abu Salim prison, with members of his cell. On September 11th, 2012, the same cell members from Khazmi's Ansar Allah terrorist cell in 2006 attacked the U.S. Consulate with him. The cell members included Egyptian nationals Asmi Ahmed, Karim Moawad al-Rahmani, Ahmed Hazaa, and Mohammad Jaber Abd al-Maqsoud.

In 2012, Khazmi was the Leader of the May 28th militia. He fought in the 2012 Bani Walid uprising from January 23rd to 25th, 2012, as the city wanted its governing authority separate from the National Transitional Council (NTC), Libya's then-acting Government. The May 28th militia was stationed in the al-Sadadah area and was fighting in support of the NTC.

Throughout his long jihadist career, he had been to Sudan, Egypt, Syria, and Iraq and had maintained close associations with terrorists in each locale who remained loyal to him. In 2013, he moved to Sirte and joined AAS-S with a number of fellow detainees from Abu Salim prison. At the time, he received material support from the Libyan Government in Tripoli for AAS-S. Khazmi then moved to Darnah and joined ISIS, taking a senior leadership position with the group. He issued ISIS assassination orders and was involved in several terrorist activities in Sirte, Zliten, and Darnah.

On September 27th, 2019, Khazmi was killed in a U.S. Military airstrike with fellow Benghazi attacker Mahmoud al-Barassi. They had been targeted at a gathering of ISIS members at a facility that an Indian company abandoned during the revolution. Again, it was not reported at the time that the attack killed U.S. Consulate in Benghazi attackers. A funeral for Khazmi was held inside his family home in Bani Walid.

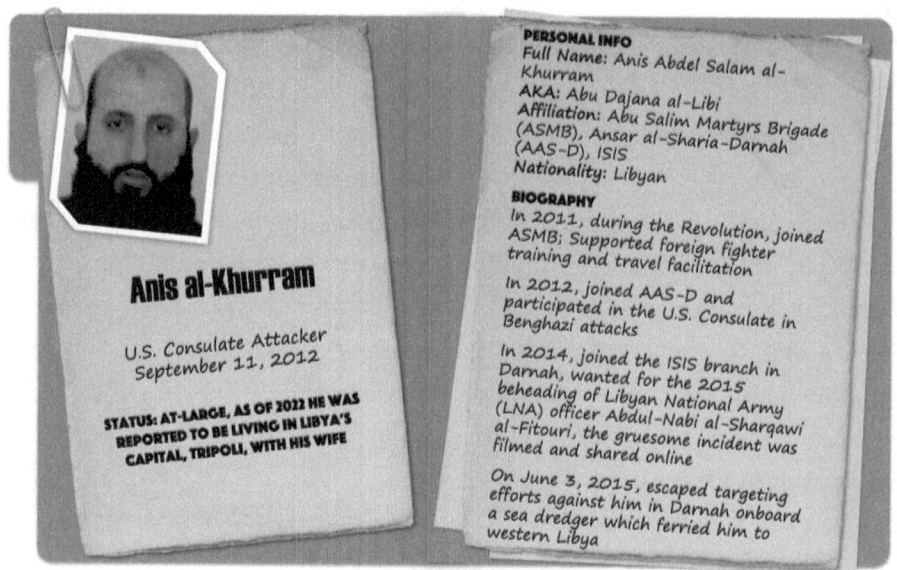

(78) Anis al-Khurram, full name Anis Abdel Salam al-Khurram with alias Abu Dajana al-Libi from the island of Azzouz in Darnah. In 2011, during the Libyan revolution, Anis was a member of the Abu Salim Martyrs Brigade (ASMB). He initially was a document forger for the group, providing documents to foreign terrorists for easy travel in and out of Libya. Further, he went on to oversee training camps for Libyans and foreigners, and then would assist in facilitating the trainees to other conflict zones.

In 2012, Anis joined AAS-D. On September 11th, 2012, he traveled with several AAS-D terrorists to Benghazi to participate in the U.S. Consulate attacks. In 2014, he pledged allegiance to the ISIS branch in the city of Darnah and was a senior leader in the local group. He was on "Most Wanted" lists in Libya as he was linked to several terrorists and terrorist operations besides the Benghazi attacks. For example, in 2015, he reportedly carried out the beheading of LNA officer Abdul-Nabi al-Sharqawi al-Fitouri in the courtyard of an ancient mosque in Darnah. The gruesome video was posted online by ISIS, and they allowed children to pose while holding the slain officer's head.

On June 3rd, 2015, Anis escaped targeting efforts against him in

Darnah by the LNA onboard a sea dredger enroute to western Libya. While he had moved throughout western Libya between Tripoli, Sabratha, and Zawiya, as of 2022, he was reported to be living in Tripoli with his wife.

(79) Alaa al-Kilani, full name Alaa el-Din Salem Hassan al-Kilani and variant Alaa Salem Hassan al-Shukri from the al-Laythi neighborhood, Benghazi. Alla was arrested on terrorism charges in 2007 for supporting global terrorist facilitation networks to include al-Qa'ida. He served in Abu Salim prison. In 2011, during the revolution, he fought with his neighbor Ahmed Abu Khatallah in Khatallah's neighborhood militia Ubaydah bin Jarrah (UBJ). Alaa was involved in many assassinations in Benghazi, including the assassination on July 28th, 2012, of Colonel Suleiman Bourziza, one of the Libyan Military Intelligence leaders in Benghazi.

On September 11th, 2012, he participated in the attack on the U.S. Consulate in Benghazi with his brother AQIM member Hassan al-Shukri. Alaa fought against the LNA in the Battle of Benghazi, which lasted from 2014 to 2017. He then fled to Tripoli and was arrested in early 2022.

Enemies: Brothers in Arms—The Attackers

(80) **Salem al-Lawati**, full name Salem Yassin Ali al-Lawati with alias Ahmed Yassin from Benghazi. In 2007, Salem was arrested and charged with terrorism for supporting facilitation networks aiding global terrorist networks to include al-Qa'ida. He was sent to Abu Salim Prison. In 2011, during the Libyan revolution, he joined the Rafallah al-Sahati Brigade.

In 2012, he joined AAS-B. He was present in Benghazi during the first annual Sharia conference in June 2012, hosted alongside many regional terrorist groups and Islamist militias. Several months later, in September 2012, he participated in the U.S. Consulate in Benghazi attacks with AAS-B and terrorist Ahmed al-Fitouri's group. After the attacks, Salem participated in the September 27th, 2012, attack on the Benghazi Security Directorate.

The status of Salem after September 2012 was unknown. While several of the Consulate attackers fled to Syria, there was no confirmation of him ever being located in Syria.

M

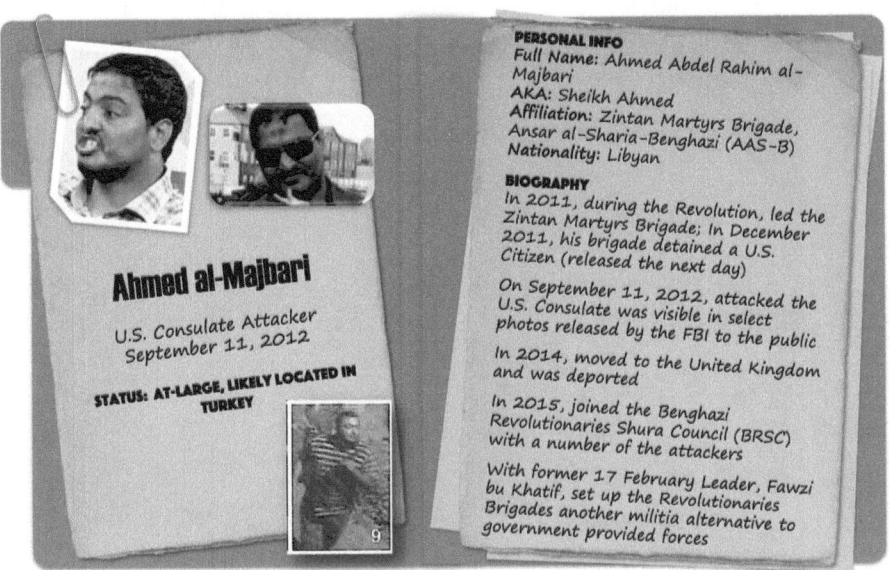

Ahmed al-Majbari
U.S. Consulate Attacker
September 11, 2012

STATUS: AT-LARGE, LIKELY LOCATED IN TURKEY

PERSONAL INFO
Full Name: Ahmed Abdel Rahim al-Majbari
AKA: Sheikh Ahmed
Affiliation: Zintan Martyrs Brigade, Ansar al-Sharia-Benghazi (AAS-B)
Nationality: Libyan

BIOGRAPHY
In 2011, during the Revolution, led the Zintan Martyrs Brigade; In December 2011, his brigade detained a U.S. Citizen (released the next day)

On September 11, 2012, attacked the U.S. Consulate was visible in select photos released by the FBI to the public

In 2014, moved to the United Kingdom and was deported

In 2015, joined the Benghazi Revolutionaries Shura Council (BRSC) with a number of the attackers

With former 17 February Leader, Fawzi bu Khatif, set up the Revolutionaries Brigades another militia alternative to government provided forces

(81) **Ahmed al-Majbari**, full name Ahmed Abdel Rahim al-Majbari with alias Sheikh Ahmed from the al-Laythi district in Benghazi. During the Libyan revolution in 2011, he led the Zintan Martyrs Brigade in Benghazi. This is a separate group from the Zintanis in Tripoli. On December 1st, 2011, this Brigade detained a U.S. Citizen NGO representative after raiding a compound he was at. The Brigade reported that they believed persons in the compound were Gaddafi supporters. The U.S. Citizen and 21 other individuals were released a day later. Additionally in 2011, Majbari was an accused murderer to include a killing at the Ajdabiya Hospital. From 2011 to 2014, Majbari was funded by the Government of Libya.

During the September 11th, 2012 attacks, Majbari was visible in a small subset of attacker photographs shared by the FBI with the public in 2013. After the attacks, during the Second Libyan Civil War, he led Zintan Martyrs Brigade terrorists to attack LNA camps in Benghazi, especially the Saiqa camps. In 2014, he joined the BRSC with a large number of the attackers. Majbari and Fawzi bu Khatif (17 February's former Leader) also established the Revolutionaries Brigade, which again was set up as a militia alternative to the military and other security forces like the police—similar to 17 February in the past.

After the Benghazi attacks, Majbari moved to the United Kingdom after his wife was given a scholarship in 2014. There was photographic evidence of him located in Manchester, United Kingdom. The British government ended up deporting him and the terrorist Adel Balaam. In later years, Majbari moved to Turkey and became a resident, where he was at-large as of 2022.

(82) Muharib al-Majbari, full name Muharib Musa Mohammad al-Majbari with alias Al Dabour (The Hornet) from Benghazi. Muharib was a former detainee at Abu Salim prison and was released before the Libyan revolution. In 2011, during the revolution, Muharib joined 17 February and was closely associated with leaders Ismail al-Sallabi, Alaa al-Ramli, and Khaled Balam. While Muharib was a bona fide fighter, he also played a professional role as a documentarian covering terrorist exploits in Benghazi.

On September 11th, 2012, Muharib participated in the attacks on the U.S. Consulate in Benghazi. In 2012, Muharib also formed an assassination squad with terrorists Zakaria al-Zawi, Mohammad al-Alam al-Alwani, and Ihab al-Jazawi. The squad specialized in assassinating military officers and security officials in the al-Laythi neighborhood. For example, on July 26th, 2013, Muharib, Zakaria, Mohammad, and Ihab executed Brigadier General Salem al-Sarrah while he was praying in the al-Tawbah Mosque in the al-Laythi area. Muharib was at-large after he fled Benghazi to western Libya. As of 2022, he was located in Tripoli, where he promotes extremism but provides religious lessons. Zakaria was detained in Spring 2022, and Mohammad was captured and sentenced to death. Ihad was also at-large.

As an aside, as a large number of attackers were involved in these assassination cells, it is important to note that al-Qa'ida Senior Operative Abu Anas al-Libi, real name Nazih Abdul-Hamed Nabih al-Ruqai'i, was the actual mastermind behind the whole assassination platform in Benghazi, and he helped set it up starting in 2011. Key collaborators helping to align local groups to support assassinations were Benghazi attacker Ramadan al-Rubaie and terrorist Alaa al-Ramli. The assassinations against the police, the military, judges, and security officials carried out throughout the city of Benghazi from 2011 until the 2013 capture of Abu Anas were reported back to him, with many performed at the behest of al-Qa'ida. When we were in Benghazi, these assassinations terrorized the city, and it felt like there was at least one assassination a day.

(83) Jalal Makhzoum, full name Jalal Ahmed Mohammad Makhzoum with alias Halquma, from Benghazi. Jalal was a member of al-Qa'ida. In 2001, during the Libyan revolution, Jalal fought in Benghazi and called his militia Saraya Sufyan al-Thawri. The name was because he graduated from the retreat coined Sufyan al-Thawri at one of the al-Salmani mosques.

At the end of the revolution, Jalal joined Rafallah al-Sahati Brigade. He then decided to move on from the Brigade and establish a branch of al-Qa'ida's al-Farouq Brigade, also referred to as Saraya al-Farouq in Benghazi. As such, Jalal became the leader of the al-Farouq Brigade in Benghazi, and then al-Qa'ida had al-Farouq Brigades in additional cities with separate leaders to include Bashir al-Faki and Ali al-Tir. Those additional cities also included Misrata, Sirte, and al-Zawiya. Jalal then, in 2012, aligned the al-Farouq Brigade in Benghazi under AAS-B's umbrella; however, he remained the Brigade's Leader.

Enemies: Brothers in Arms—The Attackers

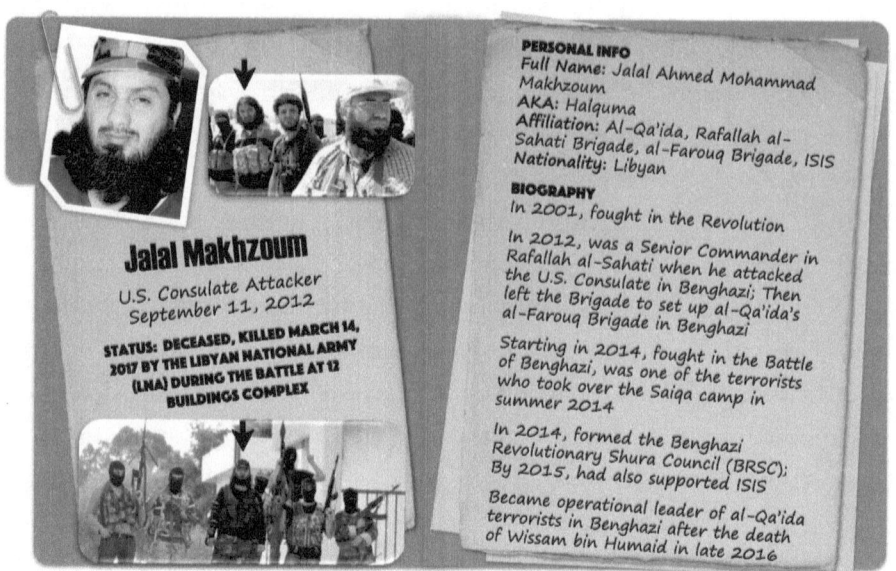

Jalal Makhzoum
U.S. Consulate Attacker
September 11, 2012

STATUS: DECEASED, KILLED MARCH 14, 2017 BY THE LIBYAN NATIONAL ARMY (LNA) DURING THE BATTLE AT 12 BUILDINGS COMPLEX

PERSONAL INFO
Full Name: Jalal Ahmed Mohammad Makhzoum
AKA: Halquma
Affiliation: Al-Qa'ida, Rafallah al-Sahati Brigade, al-Farouq Brigade, ISIS
Nationality: Libyan

BIOGRAPHY
In 2001, fought in the Revolution

In 2012, was a Senior Commander in Rafallah al-Sahati when he attacked the U.S. Consulate in Benghazi; Then left the Brigade to set up al-Qa'ida's al-Farouq Brigade in Benghazi

Starting in 2014, fought in the Battle of Benghazi, was one of the terrorists who took over the Saiqa camp in summer 2014

In 2014, formed the Benghazi Revolutionary Shura Council (BRSC); By 2015, had also supported ISIS

Became operational leader of al-Qa'ida terrorists in Benghazi after the death of Wissam bin Humaid in late 2016

In 2012, Jalal attended AAS-B's first annual Sharia conference in Benghazi. On September 11th, 2012, he participated in the attacks on the U.S. Consulate in Benghazi. Like AAS-B Leader Zahawi, Jalal was quite public about his affiliations to al-Qa'ida, unlike Jalal's close associate Mohammad al-Gharabi, who, while the most senior al-Qa'ida operative in Benghazi at the time, one would have known no association. On September 27th, 2012, he participated in the attack on the Benghazi Security Directorate.

Starting in 2014, Jalal fought with the terrorists in the Battle of Benghazi against the LNA. In summer 2014, he was seen fighting on video when terrorists overthrew the Saiqa camp in Benghazi with close allies Zahawi and Wissam bin Humaid. That year, he also became the Administrative Official of the al-Qa'ida-affiliated BRSC after its formation. By 2015, he joined ISIS in Benghazi, posting a video showcasing him as a group member in September 2015.

During the Battles of Benghazi, he was wounded in both hands and traveled to Turkey to seek treatment. He was able to continue fighting and returned to Benghazi. In December 2016, after the death of Wissam bin Humaid, Jalal became al-Qa'ida's Military Leader. He was charged with defending the group's little remaining territory in Benghazi as al-Qa'ida's base in Ganfouda essentially fell with the death of Wissam.

While Ganfouda was essentially the last stand for terrorists in Benghazi, as of February 2017, the terrorist were still fighting for primacy over one last stronghold, the 12 Buildings neighborhood in the western axis of Benghazi. This location was a bit complicated for the LNA at the time, as the terrorists had moved prisoners they kidnapped from the LNA's military police and were using those persons as human shields. During the battle at the 12 Buildings, Jalal appeared on video injured after just having survived a sniper shot from an LNA officer. It was not long after when the LNA finally killed Jalal during its shelling of the abandoned complex on March 14th, 2017. Former AQI terrorist Mohammed el-Dresi publicly confirmed and then mourned Jalal's death on March 16th, 2017. Jalal was eulogized by AQAP in the Al-Masra newspaper, by the BRSC, and by the Rafallah al-Sahati Brigade.

Next, an incident occurred that has been widely reported via the press. Whereas five days after Makhzoum's death and burial, his body was exhumed by Mahmoud al-Werfalli, who was a Libyan General in the Saiqa Brigade affiliated with LNA. Mahmoud had been charged with war crimes for the incident. While the story only gets told focused on the last day, let's also explain it starting from the first day as nothing is perfectly black and white in a warzone.

Jalal with al-Farouq brigade kidnapped an LNA soldier from the Thunderbolt Battalion, then brutally tortured him to death on video, and used his deceased body in a suicide bombing in Benghazi (please note this was his modus operandi, it was common for him to horrifically murder LNA soldiers). He then took the video and sent it to the mother of this LNA officer. The mother of the slain officer then contacted Mahmoud asking for him to investigate the incident and to bring terrorist Jalal to justice. When Jalal was finally killed in battle and it was confirmed via Mohammed el-Dresi's announcement, Mahmoud visited the mother to tell her he was finally dead. The mother said she would not believe he was dead if she did not see it with her own two eyes. So Mahmoud exhumed Jalal's body and took photographs to prove it to the grieving mother.

Enemies: Brothers in Arms—The Attackers

(84) Marei al-Manfi, full name Marei Mohammed Hassan al-Manfi from Benghazi. In the mid-2000s, Marei joined EIJ. In 2007, he was detained in Abu Salim prison for supporting global terrorist networks along with several of the Consulate attackers. In 2011, during the Libyan revolution, Marei fought with the Rafallah al-Sahati Brigade.

In 2012, he joined AAS-B and participated in the attacks on the U.S. Consulate in Benghazi. He led assassinations against military and security officers in Benghazi's Majouri and al-Hadayek neighborhoods. He reported to Benghazi attacker Ali al-Karshini in Majouri. In 2014, Marei fought with terrorists against the LNA in the Battles of Benghazi and was also associated with ISIS and fought with them in Benghazi.

In mid-April 2015, Marei publicly threatened that he would make mortars rain on the residents of Majouri and began to shell the neighborhood shortly after announcing his intentions. This led LNA to bomb Marei's home in late April 2015. At first, he was reported to have fled to Tripoli; however, in August 2015, LNA killed him in the Safsafa project area in the Qawarsha neighborhood of Benghazi.

Mohammad al-Manfi

U.S. Consulate Attacker
September 11, 2012

STATUS: DECEASED, KILLED BY THE LIBYAN NATIONAL ARMY (LNA) IN THE BATTLE OF BENGHAZI

PERSONAL INFO
Full Name: Mohammad Abd al-Salam al-Manfi
Affiliation: Martyrs Group
Nationality: Libyan

BIOGRAPHY
Prior to 2011, detainee at Abu Salim prison; Freed at start of the Revolution

In 2011, traveled to Misrata, Libya to fight against Gaddafi forces with Mohammad al-Zahawi's fighting unit which included fellow Benghazi attackers Ahmed al-Mushaiti, Hamdi Abbasi and Ali al-Karshini

Zahawi's fighting unit was under the leadership of the terrorist Bashir al-Faqi, who fought with al-Qa'ida in Afghanistan and had relationships with Abu Sayyaf and the Taliban

In approximately 2014, pledged allegiance to ISIS and joined the al-Ansar Mosque Cell in Sirte which was the ISIS epicenter in Libya at the time

Starting in 2014, fought against the LNA in the Benghazi

(85) **Mohammad al-Manfi**, full name Mohammad Abd al-Salam al-Manfi from Benghazi. Manfi was a Martyrs Group member and a former detainee in Abu Salim prison. In 2011, he was released at the start of the revolution. After leaving prison, Manfi went to Misrata to fight against Gaddafi's forces with Mohammad al-Zahawi's fighting unit. In addition to Zahawi and Manfi, the unit included fellow Benghazi attackers Ahmed al-Mushaiti, Hamdi Abbasi, and Ali al-Karshini. Another high-profile terrorist Mohammad al-Ferjani was also involved; however, he died before the Benghazi attacks. Of note, Zahawi's fighting group was under the leadership of the terrorist Bashir al-Faqi, who fought with al-Qa'ida in Afghanistan and had direct relationships with Abu Sayyaf and the Taliban in the past.

On September 11th, 2012, Manfi attacked the U.S. Consulate with terrorists from Zahawi's fighting unit from Misrata. In approximately 2014, Manfi pledged allegiance to ISIS and joined the al-Ansar Mosque Cell. At one point, Manfi reported via his Facebook page the death of German terrorist Denis Cuspert with aliases Deso Dogg and Abu Talha al-Almani. As Deso Dogg was reported to have died in 2018, this may have been erroneous information at the time of posting.

Enemies: Brothers in Arms—The Attackers

Deso Dogg was directly associated with three German terrorists who attacked the U.S. Consulate as they had been fellow members of the German-based Millatu Ibrahim group. Deso Dogg traveled into Cairo, Egypt, with a grouping of terrorists using the rouse that they were tourists, then fled to the Marsa Matrouh area in Egypt. He had still been in training camp on the Egyptian side of the border when the attacks occurred. The other terrorists from Millatu Ibrahim were in Darnah receiving indoctrination training, so were asked to support al-Qa'ida in the attacks. Manfi was killed fighting the LNA in the Battle of Benghazi.

(86) Mohammad Jaber Abd al-Maqsoud, full name Mohammad Jaber Abd al-Wahhab Abd al-Maqsoud with alias Yasser from Egypt. In 2006, Mohammad was arrested with terrorist cell members Karim Moawad al-Rahmani, Asmi Ahmed, and Ahmed Hazaa. All belonged to Malik al-Khazmi's Ansar Allah-affiliated cell. The cell planned to travel to Iraq via Syria to fight the U.S. forces with AQI.

In 2011, after Mohammad escaped from Abu Salim prison at the start of the revolution, he traveled to Benghazi and joined terrorists in the Qawarsha neighborhood. In 2012, he joined AAS-B, and he also joined al-Qa'ida in the Islamic Maghreb's (AQIM) Mali Group. On September

11th, 2012, Mohammad attacked the U.S. Consulate in Benghazi with his 2006 cell members Malik, Karim, Asmi, and Ahmed. From 2014 to 2017, he fought with terrorists in the Battle of Benghazi against the LNA. After LNA won the war, Mohammad fled to the desert and went dark. As of 2022, it was believed that Mohammad reintegrated into society unnoticed and may no longer be involved in terrorist activities.

Hamza al-Darnawi

U.S. Consulate Attacker
September 11, 2012

STATUS: AT-LARGE, LIKELY LOCATED IN LIBYA

PERSONAL INFO
Full Name: Abd al-Hamid Mohammad Abd al-Hamid Masli
Affiliation: Al-Qa'ida
Nationality: Libyan

BIOGRAPHY
In 2003, departed for Afghanistan via Iran to join al-Qa'ida and to fight U.S. forces

Closely associated with Osama bin Laden, was the leader and trainer of an al-Qa'ida electronics and explosives workshop producing improvised explosive device (IED) components

After the Libyan Revolution in 2011, Hamza returned to Darnah where he trained al-Qa'ida terrorists in a number of regional camps in support of long-time al-Qa'ida operative Abdul Basit Azzouz

Hamza traveled with some of his trainees to Benghazi to participate in the Consulate attack on September 11, 2012

(87) Hamza al-Darnawi, real name Abd al-Hamid Mohammad Abd al-Hamid Masli from Darnah. Hamza was a member of core al-Qa'ida. In 2003, he departed Libya to travel to Afghanistan to join al-Qa'ida and fight U.S. forces. He first traveled through Syria and then Iran, where he met up with fellow Libyan terrorist Abdul Ghaffar al-Tashani in Iran. He then traveled to Waziristan in Pakistan's Federally Administered Tribal Areas (FATA).

Initially, Hamza received training from al-Qa'ida at terrorist camps in the FATA, but he rose in the ranks and monitored the new trainees. Trainees included a mix of foreign fighters from Britain, Syria, and Iraq. During these early years, he formed a close working relationship with then al-Qa'ida Leader Osama bin Laden due to Hamza being a specialist in the manufacturing of explosive devices using detonation mechanisms via remote locations. As Bin Laden searched for new and innovative ways to attack western targets, he leaned on Hamza as an expert in the organization. To this end, Bin Laden supplied Hamza with his own electronics and explosives workshop.

After the Libyan revolution in 2011, Hamza traveled to Libya through the Turkey-Sudan terrorist pipeline. He returned to Darnah, where he trained al-Qa'ida terrorists in several regional terrorist camps. He assisted long-time al-Qa'ida operative Abdul Basit Azzouz in training the fighters he recruited for al-Qa'ida, which numbered about 200 terrorists in 2011. However, the Turkey-Sudan terrorist pipeline reportedly brought in approximately 1000 more al-Qa'ida terrorists for training in 2011.

Starting in 2012, Azzouz focused on preparing his trainees for Syria. He was also involved in several other Syria-related efforts, including sending weapons and humanitarian aid to terrorists fighting primarily for the al-Nusrah Front at the time. On September 11th, 2012, Hamza traveled with some of his al-Qa'ida trainees to Benghazi to participate in the U.S. Consulate attacks.

After the Benghazi attacks, Hamza reportedly stayed in Libya. As he was known to issue death threats to prevent others from taking photographs of him, there were no currently available images of him—the only one available, is now 20 years old. As of 2022, he was at-large. In 2014, Azzouz reportedly got snatched off the street in Turkey likely by Turkish authorities, whereabouts unknown.

(88) Ali Medlej was born in Beirut, Lebanon, and in 1998 migrated to Canada. Ali was a member of AQIM and fought in the battalion led by U.S. Consulate in Benghazi attacks Mastermind MBM called the al-Mulathameen Battalion.

In the late 2000s, he was radicalized in Canada with fellow Benghazi attacker Xris Katsiroubus. In 2011, he left Canada with Xris to attend Islamic School in Mauritania. In late 2011, Medlej was arrested during a raid in Nouakchott that wrapped up 36 suspected al-Qa'ida terrorists. He was interrogated for about 40 days and then released. In early 2012, he departed Mauritania for Algeria, where he found work at the In Amenas

Oil Facility in Algeria and gathered intelligence against the facility and its staff.

On September 11th, 2012, Ali, Xris, and Abu Bara al-Jazairi attacked the U.S. Consulate in Benghazi. All three were chosen to participate in the attacks due to their command of English, as the Consulate attack was a planned kidnapping operation. They were to be used to hold the Ambassador hostage while AQIM negotiated for terrorist prisoner releases.

In January 2013, Ali's prior insider threat casing activities enabled a successful AQIM terrorist operation. Starting on January 16th, 2013, Ali, Xris, and Abu Bara carried out the three-day attack and hostage crisis at the In Amenas oil facility that killed 37 hostages from 9 different countries, including 3 Americans. The attack also killed one local Algerian guard based at the facility. Both attackers Ali and Xris were killed, as well.

(89) Asmi Ahmed, full name Asmi Ahmed Mohamedin from Egypt. In 2004, Asmi led a terrorist cell in Egypt affiliated with Ansar Allah with the following terrorists Hani Mohammad Abdel Magoud with alias Ibrahim; Bahaa Ali Ali Abu al-Maati with alias Adham; and fellow Benghazi attackers Mohammad Jaber Abdel Wahab, Karim Moawad al-Rahmani, and Ahmed Hazaa. The cell had pledged allegiance to the terrorist Asmi Ahmed Amira (a separate individual from subject).

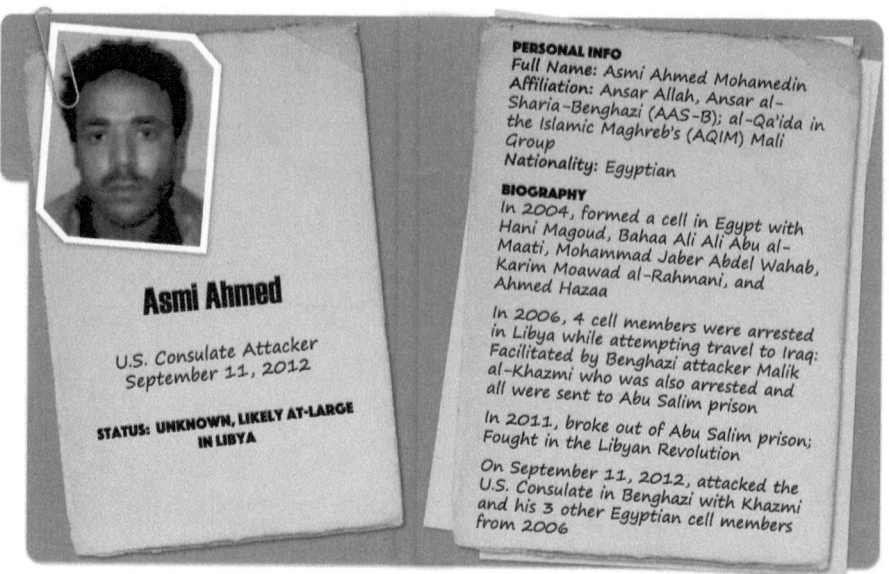

Asmi Ahmed

U.S. Consulate Attacker
September 11, 2012

STATUS: UNKNOWN, LIKELY AT-LARGE IN LIBYA

PERSONAL INFO
Full Name: Asmi Ahmed Mohamedin
Affiliation: Ansar Allah, Ansar al-Sharia-Benghazi (AAS-B); al-Qa'ida in the Islamic Maghreb's (AQIM) Mali Group
Nationality: Egyptian

BIOGRAPHY
In 2004, formed a cell in Egypt with Hani Magoud, Bahaa Ali Ali Abu al-Maati, Mohammad Jaber Abdel Wahab, Karim Moawad al-Rahmani, and Ahmed Hazaa

In 2006, 4 cell members were arrested in Libya while attempting travel to Iraq: Facilitated by Benghazi attacker Malik al-Khazmi who was also arrested and all were sent to Abu Salim prison

In 2011, broke out of Abu Salim prison; Fought in the Libyan Revolution

On September 11, 2012, attacked the U.S. Consulate in Benghazi with Khazmi and his 3 other Egyptian cell members from 2006

All the cell members traveled to Libya in the years that followed. In 2006, Asmi and the three cell members (who later participated in the Benghazi attacks) traveled to Libya to enter the foreign fighter facilitation pipeline to fight with AQI. In Libya, they connected with Ansar Allah cell leader Malik al-Khazmi, and all five were arrested on terrorism charges and sent to Abu Salim prison. In addition to Khazmi, the Egyptian cell was connected with Libyan terrorists Miftah Zaltoum, Mubarak Ibrahim Abu Bakr Amer, and his brother Ismail Ibrahim Abu Bakr Amer.

In 2011, Asmi and the rest of the cell were freed from Abu Salim prison at the start of the revolution. In 2012, he joined AAS-B and AQIM's Mali Group. On September 11, 2012, he carried out the attacks on the U.S. Consulate in Benghazi with Khazmi and their three additional cell members from 2006.

As of 2022, the original Egyptian cell members were all likely at-large, except for Bahaa Ali, who was arrested by the LNA in Darnah on October 8th, 2018, with Egyptian terrorist Hesham Ashmawy and Benghazi attacker Marei Zoghbi. Bahaa Ali and Heshawy were both close associates of U.S. Consulate attack Mastermind MBM and helped co-found al-Mourabitoun in Darnah with MBM. As both were high-value terrorists in Egypt, LNA extradited them to Egypt, where they were executed. While there is limited reporting on Benghazi attacker, Asmi Ahmed, if he was more senior to Bahaa Ali in 2006, he would presumably be another high-value terrorist target of the Egyptian authorities. As of 2022, the status of Asmi was unknown, but he was suspected of being at-large in Libya.

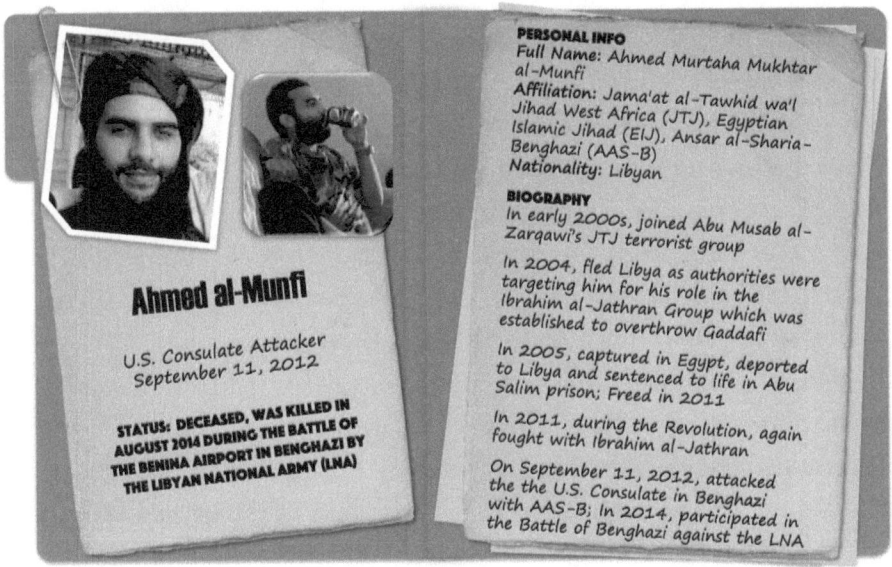

(90) Ahmed al-Munfi, full name Ahmed Murtaha Mukhtar al-Munfi with alias Abu Usayd was born in eastern Libya or the Marsa Matrouh area in Egypt. Ahmed was a member of the Jama'at al-Tawhid wa'l Jihad (JTJ), founded by Abu Musab al-Zarqawi and was also a member of the EIJ. In 2004, Ahmed fled Libya to Egypt as authorities targeted him for his membership in the Ibrahim al-Jathran Group. He was captured in Egypt and deported back to Libya.

In 2005, Ahmed was sentenced to life in Abu Salim prison along with Ibrahim al-Jathran. Jathran's terrorist group was formed to use terrorism to overthrow Gaddafi and the Libyan Government. According to western media, Jathran was in jail for "car theft". He broke out of Abu Salim prison three days after the Libyan revolution kicked off and was not released. In 2011, during the Libyan revolution, Ahmed again fought with the Ibrahim al-Jadran group. When we were in Libya in 2012, Jathran was the lead commander of an extremist militia that branded itself as an internal security force coined Petroleum Defense Guards.

In 2012, Ahmed supported the formation of AAS-B with fellow Benghazi attackers Zahawi, Mahmoud al-Barassi, Ali al-Karshini, and the al-Faydi family. Ahmed was the key official over the Guard Land neighborhood in Benghazi for AAS-B. As he comes from a family of jihadists, his terrorist brothers helped him control the neighborhood through force. His brother Mahmoud was a member of al-Qa'ida and died in a suicide bombing in Benghazi in 1997. Three other brothers were sent to Abu Salim prison; two were sentenced to life in prison, Nasser and Hatem. A third brother, Imad, was released during negotiated Libyan Government prison releases. The last brother, Khaled, operated a charitable association called al-Majd affiliated with Ansar al-Sharia in the city of al-Jufra, Libya. Al-Majd was a front organization for weapons smuggling operations where it moved weapons and ammunition to several different regional terrorist groups operating in conflict zones.

On September 11th, 2012, Ahmed participated in the attacks on the U.S. Consulate in Benghazi. In 2014, he fought in the Battle of Benghazi against the LNA, including the battles to overthrow the LNA camps. In early August 2014, Ahmed participated in terrorist attempts to take over the Benina International Airport in Benghazi. In mid-August 2014, LNA conducted a counterattack on the terrorists, and several leaders were killed, including Ahmed. Zahawi also died in the Battle of the Benina Airport.

(91) Ahmed al-Mushaiti, full name Ahmed Hassan al-Sharif Mohammed al-Mushaiti with alias Asseda from Benghazi. In 2011, during the Libyan revolution, he joined the Rafallah al-Sahati Brigade. In 2012, Ahmad joined AAS-B and was a Senior Operational Commander in the group, where he led assassinations, kidnappings, and bank robberies. On September 11th, 2012, he participated in the attack on the U.S. Consulate in Benghazi, where he was responsible for cordoning off a street parallel to the Embassy. He was manning the street with a gun truck, specifically, a Toyota truck equipped with a Soviet-built ZPU-2 14.5 anti-aircraft gun.

It was alleged that Ahmed's brother, Mohammad, worked for the Blue Mountain Group (BMG) at the Consulate, and that Mohammad attacked the Consulate on September 11th with Ahmed. BMG provided the unarmed guards at the Consulate. Such a potential pairing shows how easy it was for attackers to gain actionable intelligence on the ground, especially as this was not the only insider reported to be involved. Another member of the Mushaiti family, Benghazi attacker Salah, was also alleged to have guarded Stevens while he was the Special Envoy in 2011. The U.S. Government had long reported that BMG did not recruit from within militia ranks. That was false. A case in point is Benghazi attacker Faraj Shakku. During the revolution, he fought for 17 February, and then he was hired by BMG.

As an aside, the primary terrorist group financier in Benghazi, businessman Ashraf bin Ismail, was reportedly instrumental in getting BMG linked up with the State Department for the Consulate contract. As background, Ashraf had financed AAS-B; Ismail al-Sallabi, the al-Qa'ida Commander who was the Founder of 17 February and Rafallah al-Sahati

Brigade; Mohammad al-Gharabi, the senior al-Qa'ida commander in Benghazi and Leader of Rafallah al-Sahati Brigade; Jalal Makhzoum, a Benghazi attacker and Leader of a branch of al-Qa'ida's Malik Brigade; and Salem Darby, the Leader of Abu Salim Martyrs Brigade (ASMB) in 2012. Furthermore, when we mention attackers traveling to Turkey for treatment, while most attackers were financed for treatment and travel by the Libyan Government, Ashraf financed a number of those trips, as well. Ashraf was believed to have stolen 3 billion dollars from the Libyan Federal Reserve, using a large part of those funds to provide direct material support to terrorism.

In addition to Ashraf, Ismail al-Sallabi reportedly played a role not only in the BMG contract at the Consulate, but in the 17 February contract, which represented the armed guards at the Consulate. While Ismail and Fawzi bu Khatif were masters at hiding their involvement in the attacks, no one can ignore just how many current and former 17 February members attacked the Consulate.

In 2014, Ahmed joined ISIS and served as a Field Commander in the al-Sabri and the Souq al-Hout neighborhoods during the Battle of Benghazi. He confessed to killing LNA military officer Suleiman al-Houti in Benghazi on behalf of ISIS. This killing was videotaped and made public in February 2017. Ahmed was also involved in the ISIS executions of Egyptian expatriate workers.

In March 2017, Ahmed was arrested by the Misrata Crime Control Department after being injured by the LNA in the Qawarsha neighborhood in Benghazi and traveling to Misrata for treatment. As of August 2022, Ahmed was currently on trial in Misrata. In terms of his brother Mohammad, he was reportedly injured during the Benghazi attacks and received medical treatment in Turkey. After returning to Libya, Mohammad fled Benghazi for Misrata where he too joined ISIS. As of 2022, Mohammad's status and whereabouts were unknown.

(92) Majdi al-Mushaiti, full name Majdi al-Maliq al-Ghanai al-Mushaiti from Khamis, Libya. He was not believed to have been involved in terrorism before the Libyan revolution. Majdi, though, was born into a family of terrorists. Three separate branches of his family attacked the U.S. Consulate in Benghazi on September 11th, 2012. In 2011, he fought with the Rafallah al-Sahati Brigade. After the revolution, he bought and sold weapons and ammunition from the large stockpiles leftover in the country.

In 2012, he joined AAS-B. He participated in the Consulate attacks and was visible in the limited number of photographs released by the FBI in 2013. As a member of AAS-B, Majdi essentially took over the city center of Khamis and ran terrorism operations and assassinations out of it. In February 2014, Majdi executed seven Egyptian Coptic Christian expatriate workers near Benghazi.

Starting in the summer of 2014, Majdi fought with the terrorists in the Battle of Benghazi against the LNA. It was first reported on August 26th, 2014, that the Saiqa Forces arrested Majdi in Benghazi. Then on August 28th, 2014, he was found dead in the al-Kish area of Benghazi.

(93) Salah al-Mushaiti, full name Salah al-Din Ali Ibsekri al-Mushaiti with alias Hannibal from Benghazi. Salah was a member of the Martyrs Group. In the mid-2000s, he was in Syria and tried to sneak into Iraq to fight U.S. forces with AQI. He was stopped at the border, and as he could not get into Iraq, he left Syria and went home to Libya. In 2007, Salah was detained by the Gaddafi regime for supporting global terrorist networks. He was sent to Abu Salim prison.

Salah al-Mushaiti
U.S. Consulate Attacker
September 11, 2012
STATUS: UNKNOWN

PERSONAL INFO
Full Name: Salah al-Din Ali Absekri al-Mushaiti
AKA: Hannibal
Affiliation: Martyrs Group, Rafallah al-Sahati Brigade
Nationality: Libyan

BIOGRAPHY
In the mid 2000s, was denied entry to Iraq to join al-Qa'ida in Iraq (AQI)

In 2007, was detained by the Gaddafi regime for supporting global terrorist networks, and sent to Abu Salim prison

In 2011, joined Rafallah al-Sahati. Reportedly, may have guarded Stevens when he served as Special Envoy in Benghazi

On September 11, 2012, attacked the U.S. Consulate in Benghazi with two of his brothers, Hamza and Shoaib

Starting in 2014, participated in the Battle of Benghazi and then went completely dark

In 2011, during the Libyan revolution, he joined the Rafallah al-Sahati Brigade, a cell within 17 February. It was alleged that Salah was one of the guards performing security for Special Envoy Stevens in Benghazi in 2011. After the revolution, he became involved in local terrorist groups within the al-Laythi neighborhood in Benghazi and was involved in assassination and bombing campaigns. A number of the attackers lived in al-Laythi, and Salah was a neighbor of Ahmed Abu Khatallah as they lived in the same residential square.

On September 11th, 2012, he attacked the U.S. Consulate in Benghazi with two of his brothers, Hamza and Shoaib. Starting in 2014, he participated in the Battle of Benghazi and then went dark. As of 2022, the status of Salah was unknown.

(94) Shoaib al-Mushaiti, full name Shoaib Ali Ibsekri Shoaib al-Mushaiti from Benghazi. Shoaib got his start in terrorism with EIJ. In 2007, along with his brother Salah, he was arrested and sent to Abu Salim prison for supporting global terrorist facilitation networks to include for al-Qa'ida. In 2011, during the Libyan revolution, he joined the Ubaydah bin Jarrah (UBJ) militia, which was the militia in his neighborhood at the time. Shoaib was named in the case involving the July 28th, 2011, assassination of Major General Abdul Fatah Younis al-Obeidi, in which his brother Abdul-Fattah Ali Ibsekri al-Mushaiti was directly involved.

In 2012, he joined AAS-B and participated in the Benghazi attacks with his brothers and cousins, including Ahmed, Majdi, Hamza, and Salah. On February 10th, 2014, Shoaib's brother Abdel-Fattah was killed in an explosion at their family home. It also killed four AQIM terrorists, including Benghazi attacker Taher al-Awami. In July 2014, Shoaib was involved in overthrowing the LNA's Saiqa camp. On July 8th, 2015, he was killed by the LNA in the Battle of Benghazi.

O

(95) Abdulaziz al-Obaidi, full name Mohammad Abdulaziz al-Mahdi al-Obaidi with alias "Jandal" from Benghazi. Jandal was a member of AAS-B at the time of the attacks. He was a curator and performed the call to prayers at the al-Awzai Mosque in the al-Laythi neighborhood. In 2011, during the revolution, he joined the Rafallah al-Sahati Brigade—led in 2011 by Mohammad al-Gharabi, Ismail al-Sallabi, and Faraj al-Majbari. That same year, he appeared in a music video filmed with fellow Rafallah al-Sahati terrorist and al-Laythi neighborhood friend, Abdul Qader al-Misrati. By 2012, Abdul Qader had been in an assassination cell in Benghazi with Abdulaziz and fellow Benghazi attacker Imad al-Awami.

"Baby Hedgehogs"

Mohammed al-Obaidi

U.S. Consulate Attacker
September 11, 2012

STATUS: LIKELY DECEASED, WAS REPORTED TO HAVE BEEN KILLED BY THE LIBYAN NATIONAL ARMY AT THE END OF THE BATTLE OF BENGHAZI IN 2017

PERSONAL INFO
Full Name: Mohammad Abdulaziz al-Mahdi al-Obaidi
AKA: Jandal
Affiliation: Rafallah al-Sahati Brigade, Ansar al-Sharia-Benghazi (AAS-B), ISIS
Nationality: Libyan

BIOGRAPHY
In 2011, during the Revolution, fought with Rafallah al-Sahati

In 2012, joined AAS-B and attacked the U.S. Consulate in Benghazi with close associate Bashir Bounekhla (Bashir took the "Baby Hedgehog" selfie which trended on social media while they were fighting the LNA from the Hawari cement factory in Benghazi)

Starting in 2014, fought with the terrorists against LNA in the Battle of Benghazi

In approximately 2015, pledged allegiance to ISIS while fighting in the axis of West Benghazi

In approximately 2015, Jandal pledged allegiance to ISIS. During the time, Jandal was fighting in the axis of West Benghazi during the Second Libyan War when terrorists fought the LNA. Like the video, he participated in a selfie in front of the Benghazi cement factory during the war that became popular in Libya. The selfie, turned into a joke, with the terrorists in the image being referred to as "baby hedgehogs" among other names. The terrorist who took the selfie, Bashir Bounekhla on September 11th, 2012 attacked the U.S. Consulate with Jandal. Bashir was from Zlinten and led the assassinations of LNA officers in al-Laythi.

In 2017, it was reported that both Jandal and Bashir had been killed by LNA in the vicinity of the cement factory. However, as there was no photographic evidence that Jandal was deceased, there had been reported sightings of him in western Libya, including in Tripoli, Misrata, and Zliten.

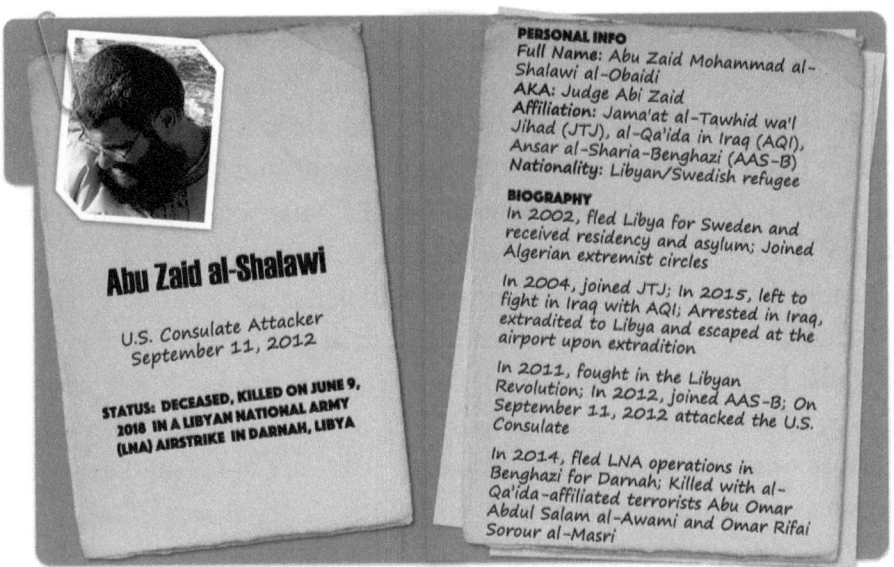

(96) Abu Zaid al-Shalawi, full name Abu Zaid Mohammad al-Shalawi al-Obaidi with alias Judge Abi Zaid from Benghazi. In 2002, he fled Libya for Sweden and received residency and asylum there. In Sweden, he became involved in Algerian extremist circles and was indoctrinated through their takfiri ideology. He joined Jama'at al-Tawhid wal-Jihad (JTJ), led by Abu Musab al-Zarqawi, which soon became AQI. In 2005, Abu Zaid left for Syria and then traveled to Iraq to join AQI and fight U.S. forces. He was detained in Iraq and then deported to Libya. Upon arrival at the airport in Tripoli, he escaped and went dark in Libya. As he had received a 15-year sentence in Abu Salim, many think he was serving in prison the entire time and broke out during the Arab Spring.

Abu Zaid was next seen during the Libyan revolution in 2011 when he advocated for Islamists. Then in 2012, he joined AAS-B and became one of the terrorist group's judges. On September 11th, 2012, he participated in the attack on the U.S. Consulate in Benghazi. In 2014, he fled Benghazi for Darnah when the LNA started counterterrorism operations in the city. In approximately 2015, Abu Zaid served as a Cultural Advisor

Enemies: Brothers in Arms—The Attackers

to the Government of Libya under Khalifa al-Ghawil, the Prime Minister of the General National Congress-led National Salvation Government in Tripoli.

On June 9th, 2018, Abu Zaid was killed when the LNA carried out an airstrike in the Shiha area of Darnah while trying to liberate Darnah from terrorists. Killed alongside Abu Zaid were al-Qa'ida in Darnah Senior Leader Abu Omar Abdul Salam al-Awami, and Omar Rifai Sorour al-Masri with alias Abu Abdullah al-Masri. Omar was the Mufti of the DMSC, an al-Qa'ida affiliate established to ensure al-Qa'ida vice ISIS controlled Darnah. Omar was also co-founder of the militant group al-Mourabitoun, alongside U.S. Consulate Mastermind MBM, Egyptian terrorist Hesham Ashmawy and fellow Benghazi attacker Marei Zoghbi. Lastly, Omar's father, Rifai Sorour, was a prominent Egyptian jihadi Salafist, who had been a close associate of former al-Qa'ida Leader Dr. Ayman al-Zawahiri in the past.

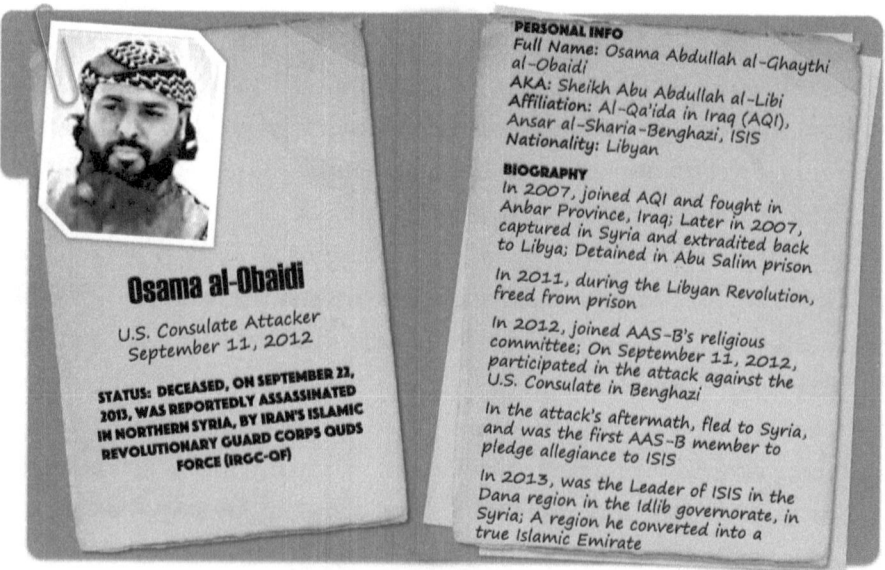

Osama al-Obaidi

U.S. Consulate Attacker
September 11, 2012

STATUS: DECEASED, ON SEPTEMBER 22, 2013, WAS REPORTEDLY ASSASSINATED IN NORTHERN SYRIA, BY IRAN'S ISLAMIC REVOLUTIONARY GUARD CORPS QUDS FORCE (IRGC-QF)

PERSONAL INFO
Full Name: Osama Abdullah al-Ghaythi al-Obaidi
AKA: Sheikh Abu Abdullah al-Libi
Affiliation: Al-Qa'ida in Iraq (AQI), Ansar al-Sharia-Benghazi, ISIS
Nationality: Libyan

BIOGRAPHY
In 2007, joined AQI and fought in Anbar Province, Iraq; Later in 2007, captured in Syria and extradited back to Libya; Detained in Abu Salim prison

In 2011, during the Libyan Revolution, freed from prison

In 2012, joined AAS-B's religious committee; On September 11, 2012, participated in the attack against the U.S. Consulate in Benghazi

In the attack's aftermath, fled to Syria, and was the first AAS-B member to pledge allegiance to ISIS

In 2013, was the Leader of ISIS in the Dana region in the Idlib governorate, in Syria; A region he converted into a true Islamic Emirate

(97) Osama al-Obaidi, full name Osama Abdullah al-Ghaythi al-Obaidi with alias Abu Abdullah al-Libi from Benghazi. Note that there are a lot of terrorists with the alias "Abu Abdullah" who are Libyan, and even one Iraqi gets confused with Osama when reviewing the ISIS network that connects back to the Benghazi attacks and the al-Battar Brigade. As such, note most biographies of "Abu Abdullah al-Libi" are incorrect as they merge several terrorists into one identity. For example, in the immediate network of the Benghazi attacks, we have three: Osama, Benghazi attacker Ahmed bin Nasser Karim, and close MBM associate and facilitator Abdullah Bukhazem, full name Al-Saadi Abdullah Ibrahim Bukhazem with alias Al-Saadi al-Nawfali. Abdullah Bukhazem was also a member of AQIM.

In 2007, Osama left Libya for Syria for onward travel to Iraq to join AQI. After arriving in Iraq, he fought with AQI in Anbar Province. During an operational visit to Syria on behalf of AQI later in 2007, Osama was captured by the Syria Regime. He was deported back to the Gaddafi regime in Libya based on a security coordination agreement between the two nations that the U.S. was involved in under then Secretary of State Colin Powell. Osama was sent to Abu Salim prison. It was not until the start of the Libyan revolution in 2011 that Osama was freed from prison.

In 2012, Osama joined AAS-B and supported several of the group's religious committee efforts. On September 11th, 2012, he participated in the attack against the U.S. Consulate in Benghazi. Soon after, he fled to Syria and was reported as the first member of AAS-B to pledge allegiance to ISIS. Again he went directly to Syria, he did not travel to Sirte, Libya, and he joined ISIS before the group established a branch in Libya. He was not a member of ISIS in Libya. He was a member of ISIS in Syria.

On September 22nd, 2013, Osama was reportedly assassinated in Bab al-Hawa, Northern Syria, by Iran's Islamic Revolutionary Guard Corps (IRGC) Quds Force. At the time of his death, Osama was the Leader of ISIS in the Dana region in the Idlib governorate in Syria. He converted the region into a true Islamic Emirate and was known to have a style where he ruled the area using Afghan Taliban-type principles.

R

(98) Karim Moawad al-Rahmani, full name Karim Moawad Mohammad al-Rahmani with alias Hamida from Egypt. On June 12th, 2006, Karim traveled from Egypt to Libya to enter the foreign fighter pipeline to Iraq to fight U.S. forces with AQI. Benghazi attacker Malik al-Khazmi was facilitating the travel for Karim through a terrorist cell he was operating in Tripoli at the time affiliated with Ansar Allah. Later in 2006, all cell members were detained and sent to Abu Salim prison.

Karim Moawad al-Rahmani

U.S. Consulate Attacker
September 11, 2012

STATUS: UNKNOWN, LIKELY AT-LARGE IN LIBYA

PERSONAL INFO
Full Name: Karim Moawad Mohammad al-Rahmani
AKA: Hamida
Affiliation: Ansar Allah, Ansar al-Sharia-Benghazi (AAS-B), al-Qa'ida in the Islamic Maghreb (AQIM)
Nationality: Egyptian

BIOGRAPHY
In 2006, arrested in Libya for belonging to an Ansar Allah terrorist cell led by Benghazi attacker Malik al-Khazmi; Intended to travel to Iraq to fight with al-Qa'ida in Iraq (AQI)

In 2011, during the Revolution, fought in Qawarsha with the al-Faydi family

In 2012, joined AQIM's Mali Group and was involved in trafficking arms to Mali and Syria

In mid 2012, joined AAS-B; On September 11, 2012, attacked the U.S. Consulate with 2006 cell members Malik al-Khazmi, Asmi Ahmed, Mohammad Jaber Abdel Wahab, and Ahmed Hazaa

At the start of the Libyan revolution in 2011, Karim escaped from Abu Salim after serving five years. He decided to stay in Libya for the revolution and was stationed in Benghazi. His zone was in the Qawarsha area, where several foreign terrorist training camps were at the time. In 2012, he joined AAS-B. He was closely associated with terrorists Faraj al-Faydi, Fawzi al-Mushaiti, and fellow Benghazi attackers Khaled al-Faydi and Younes al-Faydi.

Karim joined al-Qa'ida in the Islamic Maghreb's (AQIM) Mali Group and was involved in trafficking arms to Mali and Syria. On September 11th, 2012, he attacked the U.S. Consulate in Benghazi with Malik al-Khazmi, and other terrorists from the 2006 Ansar Allah cell, including Asmi Ahmed, Mohammad Jaber Abdel Wahab, and Ahmed Hazaa. As of 2022, the status of Karim was unknown, but he was suspected of being at-large in Libya.

(99) Ramadan al-Rubaie, full name Ramadan Mohammad al-Rubaie with aliases Kush al-Rabie and Abu Faraj al-Ansari, from Ras Obaida in Benghazi. At the time of the Benghazi attacks on September 11th, 2012, Ramadan was a senior leader in AAS-B and was the Emir of AAS-B in Ras Obaida, one of AAS-B's key strongholds in the city. He ran a military-style unit within AAS-B, which specialized in assassinations and bombings. One of his brothers, Islam, supported Ramadan in his assassination campaign. As of 2022, Islam was at-large. Ramadan had four additional brothers, like Islam, who all were involved in terrorism, which included Ahmed, Fawzi, Faraj, and Faris. Like many of the attackers, he brought at least a couple of his brothers to participate in the attack on the U.S. Consulate. His cousin, Adel al-Rubaie, participated in the attacks on the U.S. Consulate with him, as well.

Ahmed died in the Battle of Bin Jawad while fighting the Libyan revolution in 2011. Faraj and Fawzi were killed in 2016 during battles with the LNA, with Faraj dying in the Qawarsha area of Benghazi and Fawzi dying in the al-Safsfah agricultural project in western Benghazi. One of Ramadan's sons is named after Faraj. Fares, a former policeman, became a member of ISIS and reported to ISIS terrorist Mas'oud Belhassen al-Nawfali with aliases Juma'a Mas'oud al-Hassan al-Qarqa'I, Abu al-Mahdi, and Abu al-Layth. He survived an incident in al-Sabri, Benghazi, and as of 2022, was at-large in Tripoli, likely in the region near the Tripoli International Airport.

Terrorist cousin Adel was also from Benghazi. In 2012, he joined AAS-B and led assassinations targeting civilians and military personnel. Starting in 2014, he aligned himself with al-Qa'ida and fought with terrorists against the LNA in the Battle of Benghazi. He was injured by the LNA in 2017 and fled to Misrata. By 2019, he had relocated to Tripoli, fighting within the militia forces controlled by the al-Wefaq Government against the LNA. He then joined Salah Badi's al-Samoud Brigade. As of 2022, he was still at-large.

In approximately 2015, Ramadan, like Ismail, also joined ISIS. His operation was showcased in an ISIS publication titled "What did they despise when they suffered?" On January 5th, 2016, Ramadan died carrying out a suicide attack at a military gate near Ras Lanuf, Libya.

S

Yunus Emre Sakarya

U.S. Consulate Attacker
September 11, 2012

STATUS: REPORTEDLY DETAINED IN JANUARY 2019

PERSONAL INFO
AKA: Yunus al-Almani, al-Haiebi
Affiliation: Al-Qa'ida, Millatu Ibrahim, al-Nusrah Front, ISIS
Nationality: German/Turkish citizen

BIOGRAPHY
In 2008, with his brother, Ismail, arrested in Germany for an attempted plot against a local U.S. Military base; In 2010, the brothers were released

In 2012, the two traveled to Libya to train at terrorist camps; On September 11, 2012, the brothers and a German associate with alias "Abu Amina" traveled with 15 terrorists from Darnah to Benghazi to participate in the U.S. Consulate attacks; Fled Libya after the attack

Since at least 2015, was a key facilitator in the procurement of unmanned aerial vehicle (UAV) components for ISIS

He and his wife committed war crimes by enslaving a Kurdish Yazidi woman who endured abuse and sexual assault

(100) Yunus Emre Sakarya, a dual Turkish and German citizen. Yunus decided to join al-Qa'ida in October 2008 after returning to Germany from a trip to Florida. He had been mulling the idea since the 9/11 terrorist attacks in 2001. He connected with al-Qa'ida online. His first terrorist act was with his brother, Ismail, when they attempted to procure weapons with the intent to attack a U.S. Military base in Germany. As the two could not procure small arms, they decided they would instead attack the German Police, steal their weapons, and then use those weapons to attack Americans on their base. The planning was thwarted, and by the end of 2008, the two were arrested in Germany for providing material support to al-Qa'ida.

The brothers were reportedly released on parole in 2010. In 2011, they joined a new organization in Germany called Millatu Ibrahim, which was founded by Austrian terrorist Mohamed Mahmoud with alias Abu Usama al-Gharib. Gharib was the Leader of the Global Islamic Media Front (GIMF), an al-Qa'ida-affiliate, when he was arrested in 2007 for threatening to carry out attacks in Germany and Austria if the U.S. did not pull its troops out of Afghanistan.

In 2012, the Sakarya brothers traveled from Germany to Libya, transiting Turkey. Several associates from Millatu Ibrahim had their eyes set on traveling to Libya at the time including Denis Cuspert, with alias "Deso Dogg," and another Benghazi attacker that traveled with the Sakarya brothers is unidentified, but used the alias "Abu Amina." This was due to a push by German Islamist Reda Seyam for Europeans to train in Libya.

When the Sakarya brothers arrived in Libya, they trained at a camp near the Libyan border with Egypt and frequented Tobruk and Darnah. Cuspert also traveled to Libya in 2012 and shared an online post showing that the training camp was near the Mersa Matruh area in Egypt. As such, the camp may have been located on the Egypt side of the border. The Sakarya brothers and Abu Amina were in Darnah just before the attacks, receiving terrorist indoctrination lessons.

Al-Qa'ida affiliates asked the three from Germany to travel to Benghazi to participate in the attacks. It was unclear who they traveled to Benghazi with; however, it was likely the Abu Salim Martyrs Brigade. The crew left Darnah with approximately 15 terrorists. On September 11th, 2012, Yunus was tasked to block a roadway in the vicinity of the U.S. Consulate in Benghazi. After the attacks, the crew fled Libya on official Libyan passports provided by Ansar al-Sharia and processed by the Government of Libya. Most attackers that chose to flee Libya in the attack's aftermath had travel documents and passports created for them with false names before the attacks occurred, showing the level of planning involved in the al-Qa'ida plot against the U.S. Consulate.

Since at least 2015, Yunus had served as a key facilitator procuring unmanned aerial vehicle (UAV) components for ISIS. Specifically, Yunus operated an Ankara, Turkey-based front company, Profesyoneller Elektronik, to procure UAV-related equipment that totaled over half a million dollars for ISIS. In addition to Turkey, he had operated out of Mayadin, Syria, a known hub for loyalists of deceased ISIS leader Abu Bakr al-Baghdadi. As of January 2019, Yunus was reportedly detained.

In a separate issue, his wife, Jalda Sakarya, also a member of ISIS, was tried in a German court for enabling her husband's physical and sexual abuse of a Kurdish Yazidi woman that the two kept as a slave. Jalda also physically abused the woman, represented by British lawyer Amal Clooney. In July 2022, the wife was convicted of aiding and abetting genocide, guilty of crimes against humanity, and aiding and abetting crimes against humanity and war crimes. The women who marry these terrorists rarely get brought to justice when they commit many of the same crimes as their husbands. Even an FBI contract employee in its Detroit field office married terrorist Deso Dogg, and she traveled to Syria to be with him. After the fact, she was given a slap on the wrist.

(101) Hamza al-Sallak, also known as Muftah Hamza al-Sallak or Moftah el-Sallak. Limited information was available on Hamza before 2012. In 2012, he joined AAS-B and attended its first annual conference in June 2012 in Benghazi. On September 11th, 2012, he was a commander for AAS-B when he attacked the U.S. Consulate.

Starting in 2014, he fought in the Battle of Benghazi with the terrorists against the LNA. In approximately 2015, Hamza fled LNA counterterrorism operations in Benghazi and traveled to Misrata. In 2015, while in Misrata, he pledged allegiance and fought with ISIS. In 2016, as operations were intensifying against ISIS on multiple fronts from the Government of Libya, the LNA, and the U.S. through airstrikes, Hamza fled to Genoa, Italy. As of 2022, Hamza was at-large and was reported to have not returned to live in Libya. As such, the terrorist was believed to be residing in Turkey or Italy.

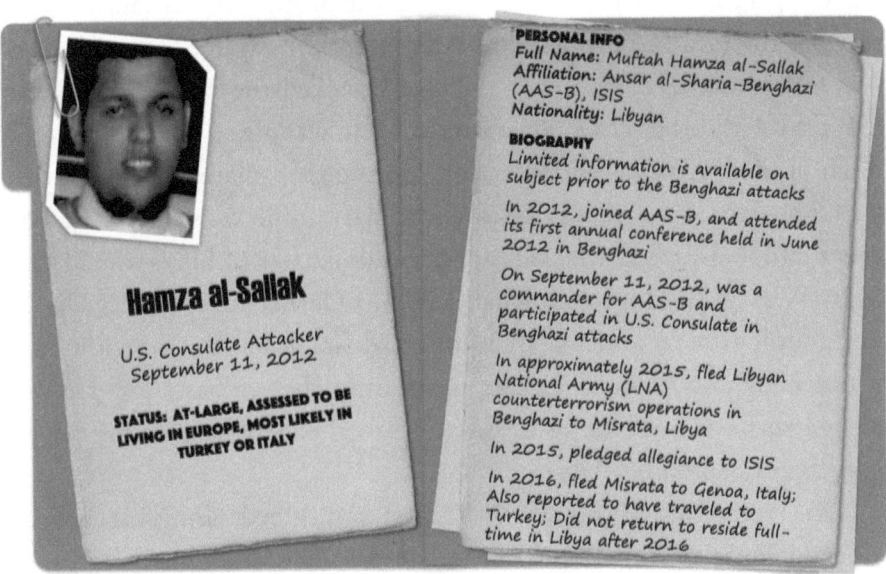

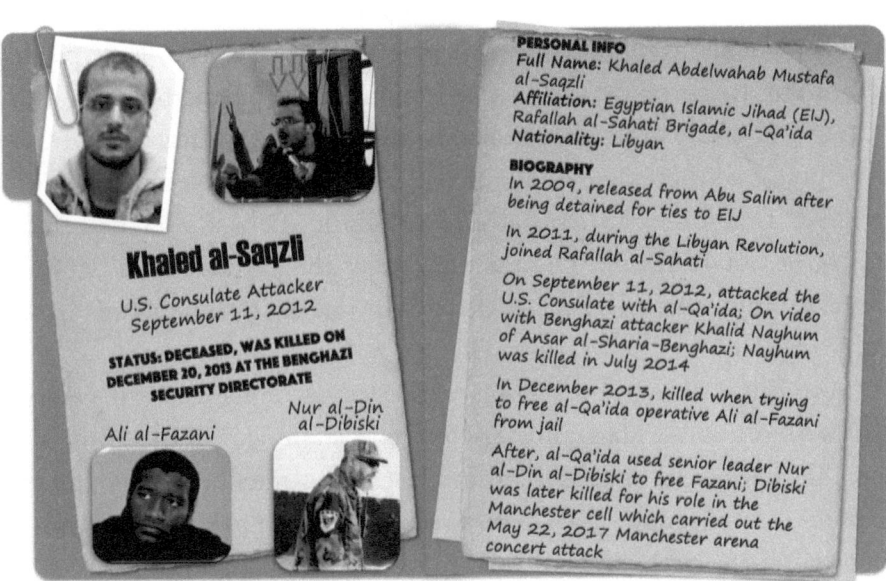

(102) Khaled al-Saqzli, full name Khaled Abdelwahab Mustafa al-Saqzli variant el-Sagezli, from Benghazi. Khaled was arrested in the past for his involvement with the EIJ and was sent to Abu Salim prison. He was released in 2009. In 2011, during the Libyan revolution, he joined the Rafallah al-Sahati Brigade and then became a member of al-Qa'ida as it had set up a base in Benghazi. Khaled was the brother of Mustafa al-Saqzli, who was the Head of the Warriors Affairs Authority after the revolution.

On September 11th, 2012, Khaled attacked the U.S. Consulate in Benghazi with other al-Qa'ida associates. He was also a close associate of Omar al-Shalaali who led the attacks on the Consulate. After departing the Consulate, Khaled was believed to have gone to seek refuge with several other al-Qa'ida attackers at the headquarters of Libya Shield One, led by Wissam bin Humaid. Khaled fled to Darnah after the attacks and stayed under the radar. Khaled's brother, Mustafa, was the individual sent to the Benghazi Medical Center to identify the Ambassador. He then assisted in having the body transported to the Benghazi airport on the morning of September 12th.

On December 20th, 2013, Khaled was killed alongside terrorist associate Abdul Karim Boghazaleh. The two died while carrying out an armed attack on the Headquarters of the Benghazi Security Directorate in an attempt to free fellow al-Qa'ida operative Ali al-Fazani, full name Ali Othman al-Fazani. Fazani was arrested a few days prior while placing a bomb under an officer's car in the Majouri area of Benghazi.

After Khaled's death, it appears Fazani had been transferred to a prison in Tripoli, and then a senior al-Qa'ida leader played a role in having him released. That leader was Nur al-Din al-Dibiski. Dibiski then had Fazani smuggled out of Tripoli. Dibiski was one of the key al-Qa'ida figures in western Libya, but we did not delve into the western networks in this publication.

As a quick background, Dibiski traveled to Afghanistan in the early 1990s to join al-Qa'ida and receive military training. Dibiski also joined the LIFG while in Afghanistan and became a member of its Military Committee. In August 2005, he was a senior leader of LIFG in Iran and was reportedly arrested by the Iranians. At some point, he relocated to Sudan, where he reportedly had a personal relationship with then President Omar al-Bashir and was involved in smuggling arms from Turkey to Sudan. In 2011, he left Sudan to fight in the Libyan revolution. He was killed with 12 other terrorists within 72 hours after the May 22nd, 2017, Manchester arena attack at the Ariana Grande concert that killed 22 patrons. Dibiski was reportedly one of the senior leaders of the Manchester Cell.

Mansour al-Shalaali

U.S. Consulate Attacker
September 11, 2012

STATUS: DECEASED, AAS-B REPORTED THE DEATH OF MANSOUR ON OCTOBER 25, 2015

PERSONAL INFO
Full Name: Mansour Juma Mohammad al-Shalaali
AKA: Al-Muhajir al-Darnawi, Abu Haroun
Affiliation: Armed Islamic Group (GIA), al-Qa'ida in the Islamic Maghreb (AQIM), Ansar al-Sharia-Benghazi (AAS-B)
Nationality: Libyan

BIOGRAPHY
In the early 1990s, went to Morocco to join the Salafi Jihadism movement
From 2005 to 2011, fought in the Algeria
In 2012, joined AQIM's Mali Group and joined AAS-B as a Commander; On September 11, 2012, attacked the U.S. Consulate with his brother Omar who was the Head of AQIM for Libya
One motivation for the al-Shalaali brothers to join the attack was to swap their brother, Adil, a prisoner in Iraq for the Ambassador; On August 21, 2016, Adil was executed in Iraq

(103) Mansour al-Shaalali, full name Mansour Juma Mohammad al-Shaalali with aliases Al-Muhajir al-Darnawi and Abu Haroun, from the eastern Shiha district in Darnah. In the early 1990s, when just a teen, Mansour traveled to Rabat, Morocco, to join the newly emerging Salafi Jihadism movement. From 2005 to 2011, Mansour fought in the Algerian Civil war with Armed Islamic Group (GIA), as did AQIM Leaders Abdel Malek Droukdal and MBM. In 2011, he left Algeria for Libya to fight in the revolution.

In 2012, he joined AQIM's Mali Group and joined AAS-B as a Commander. On September 11th, 2012, he attacked the U.S. Consulate with his brother Omar who was the Head of AQIM for Libya and the lead al-Qa'ida Commander on the ground at the Consulate. In addition to the family's historical relationship to al-Qa'ida, to include having five family members killed with AQI in Iraq, as noted a motivation for the

Shalaali brothers to join in the attack was the hope that their brother Adil could be released. The brothers wanted to pose a prisoner exchange to have Adil released from Iraqi prison for Ambassador Stevens. Adil was a renowned fighter in Iraq for al-Qa'ida and had been closely associated with the former Leader of AQI, Abu Musab al-Zarqawi. As there was no kidnapping, on August 21st, 2016, Adil was executed in Iraq.

Starting in 2014, Mansour was believed to have been fighting with his brothers and al-Qa'ida terrorists in the Battle of Benghazi. On October 25th, 2015, AAS-B reported the death of Mansour. AQIM eulogized him a few days later, praising him as a martyr, noting that he fought against the Algerian Army in Algeria for six years, being one of many Benghazi attackers to fight in the Algerian Civil War.

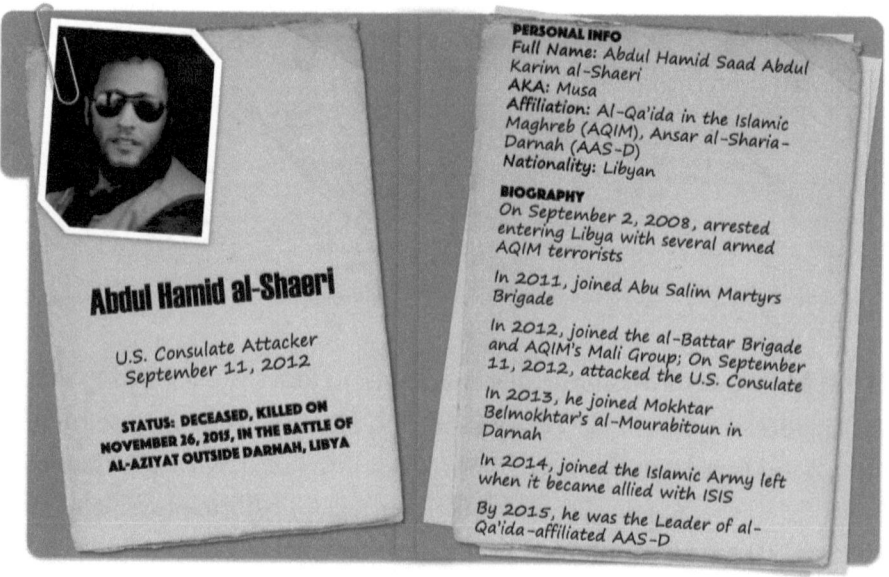

Abdul Hamid al-Shaeri

U.S. Consulate Attacker
September 11, 2012

STATUS: DECEASED, KILLED ON NOVEMBER 26, 2015, IN THE BATTLE OF AL-AZIYAT OUTSIDE DARNAH, LIBYA

PERSONAL INFO
Full Name: Abdul Hamid Saad Abdul Karim al-Shaeri
AKA: Musa
Affiliation: Al-Qa'ida in the Islamic Maghreb (AQIM), Ansar al-Sharia-Darnah (AAS-D)
Nationality: Libyan

BIOGRAPHY
On September 2, 2008, arrested entering Libya with several armed AQIM terrorists

In 2011, joined Abu Salim Martyrs Brigade

In 2012, joined the al-Battar Brigade and AQIM's Mali Group; On September 11, 2012, attacked the U.S. Consulate

In 2013, he joined Mokhtar Belmokhtar's al-Mourabitoun in Darnah

In 2014, joined the Islamic Army left when it became allied with ISIS

By 2015, he was the Leader of al-Qa'ida-affiliated AAS-D

(104) Abdul Hamid al-Shaeri, full name Abdul Hamid Saad Abdul Karim al-Shaeri with alias Musa from Darnah. Abdul was a member of AQIM. On September 2nd, 2008, he was arrested after departing Algeria to illegally enter Ghadames, Libya, with three armed terrorists from AQIM. Abdul was transferred to Tripoli and imprisoned in Abu Salim prison. On September 11th, 2012, Abdul participated in the attacks on the U.S. Consulate in Benghazi.

To run through a quick jihadist resume, at the end of 2011, after the Libyan revolution, Abdul settled back in Darnah, where he provided his terrorist expertise to many groups in just four short years. First, he joined the Abu Salim Martyrs Brigade (ASMB) in 2011. Then, he joined the al-Battar Brigade in 2012. Next, he returned to his roots in AQIM, and joined the Mali Group. In 2013, he then joined MBM's al-Mourabitoun in Darnah. Next, he took a short stint to an al-Qa'ida rival and joined the Islamic Army in 2014, and did not stay long due to his loyalty to al-Qa'ida. By 2015, he was the Leader of al-Qa'ida-affiliated AAS-D.

On November 26th, 2015, he was killed in the Battle of al-Aziyat, by the LNA. The terrorists with Abdul included Firas Abdullah al-Amami, Mohammed bin Khayal Arif al-Mansouri, Sharif al-Mansouri, Anis al-Amami, Abdul Rahman al-Haddad, Faraj al-Zobik, Monsef al-Khuram, Hamza al-Jazoi, Abdullah al-Mansoori and AQIM's Commander of the Benghazi attacks Omar al-Shalaali. At the time of his death, Shalaali was the al-Qa'ida Leader for East Libya. Both AQIM and Ansar al-Sharia eulogized Abdul after his death.

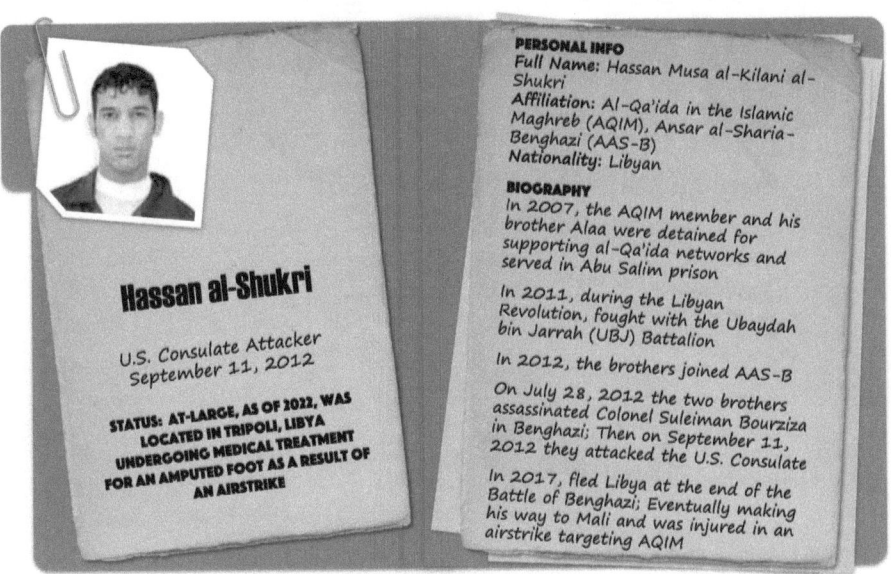

(105) Hassan al-Shukri, full name Hassan Musa al-Kilani al-Shukri from Benghazi. Hassan and his brother Alaa were detained in 2007

Enemies: Brothers in Arms—The Attackers

and served in Abu Salim prison. The brothers were released before the Libyan revolution. Hassan is a member of AQIM in its Sahara Brigades, specifically the Mali Group. In 2011, during the Libyan revolution, he fought with the Ubaydah bin Jarrah (UBJ) Battalion. In 2012, he joined AAS-B with his brother Alaa; on July 28th, 2012, the two assassinated Colonel Suleiman Bourziza in Benghazi. The two brothers then, on September 11th, 2012, attacked the U.S. Consulate in Benghazi together.

Hassan fought in the Battle of Benghazi and attempted to overthrow LNA camps with the terrorist forces. In 2017, after terrorists were defeated in Benghazi, Hassan fled to the desert to hide from follow-on LNA counterterrorism operations while his family relocated to Tripoli. In Tripoli, his brother Alaa was detained in early 2022. Hassan traveled to fight in Mali and was injured in an airstrike. He had his foot amputated, and his family brought him back to Tripoli for medical treatment. As of 2022, Hassan was based in Tripoli.

Abdul-Ati Abu Sitta

U.S. Consulate Attacker
September 11, 2012

STATUS: DECEASED, KILLED ON AUGUST 28, 2018 IN A U.S. AIRSTRIKE IN BANI WALID, LIBYA (SEE IMAGE BELOW OF THE VEHICLE HE WAS TRAVELING IN)

PERSONAL INFO
Full Name: Abdul-Ati al-Shtiwi Abu Sitta
Affiliation: Ansar al-Sharia-Sirte (AAS-S), ISIS
Nationality: Libyan

BIOGRAPHY
Former Abu Salim prisoner; One of the founding members of AAS-S along with his brother Ali

On September 11, 2012, attacked the U.S. Consulate in Benghazi with his other brother Nouri; Then fled to Syria

Returned to Libya in 2014 to work as a coordinator between terrorists in Benghazi and Sirte, and became affiliated with ISIS (see image at Abu Musab al-Zarqawi Mosque in Sirte)

In August 2018, killed in a strike targeting ISIS's Walid Abu Hariba, formerly of al-Qa'ida in Iraq (AQI); It was never reported that a Benghazi attacker was killed; As of 2022, the status of Abdul-Ati's two brothers was unknown

(106) Abdul-Ati Abu Sitta, full name Abdul-Ati al-Shtiwi Abu Sitta, with alias Abu Muslim al-Libi from Sirte, Libya. He was a member of Ansar al-Sharia-Sirte (AAS-S) after being released from serving in Abu Salim prison. He was one of the original members of AAS-S with his brother

Ali al-Shtiwi Abu Sitta; later, another brother joined as well, Nouri al-Shtiwi Abu Sitta. He worked as a legislator for the group and helped establish it with fellow Benghazi attacker Atef al-Karami; with Wissam al-Zaidi from Benghazi (the uncle of Benghazi attacker Mohamed ben Dardaf); with Abdelhadi Zarqun from the Warfalla tribe in Sirte; with Fawzi al-'Ayat from the Hawsana tribe in Sirte; and with Ahmed Ali Al Tir from Misrata. Abdul-Ati participated in several crimes in Sirte, from assassinations to bank robberies and kidnappings.

On September 11th, 2012, Abdul-Ati and his brother Nouri attacked the U.S. Consulate in Benghazi. The brothers were related to Annex attack Mastermind Wissam bin Humaid through marriage. Similar to the pattern of many attackers, Abdul-Ati immediately fled to Syria after participating in the attacks. In Syria, he joined Jaish al-Muhajireen wal-Ansar (Muhajireen Brigade) along with fellow Benghazi attacker Adel Ahmad al-Abdali. Again the group was led at the time by Abu Talha al-Libi with real name Abdul Moneim al-Hasnawi. Hasnawi went on to be a Senior Leader in AQIM. Abu-Ati appeared in an interview with Al-Jazeera discussing an attack on a U.S. Military installation, and he was with terrorist al-Hasnawi at the time.

In 2014, Abdul-Ati returned to Libya to support terrorists in Benghazi in their war against the LNA. He was responsible for coordinating between terrorists from the BRSC and those based in Sirte. By 2015, AAS-S was primarily defunct, with many former leaders like Abdul-Ati choosing to support ISIS in Sirte. On August 28th, 2018, he was killed in a U.S. airstrike on al-Dahra road near the Shemekh area, Bani Walid, that reportedly killed Walid Abu Hariba, a member of the ISIS-affiliated Kiwi group. It was never reported that this strike killed a Benghazi attacker. As of 2022, the status of his two brothers was unknown.

(107) Emad al-Shuqabi, full name Emad Faraj Mansour al-Shuqabi from Benghazi. During the 2011 Libyan revolution, he was a member of Wissam bin Humaid's Free Libya Martyrs Brigade. He joined Libya Shield One in 2012 and was also a senior leader in the group under Wissam at the time of the attacks. In Libya Shield, Emad had two roles. One was as a military commander and the second as a financial coordinator for Wissam. It had been reported that he and Wissam were cousins.

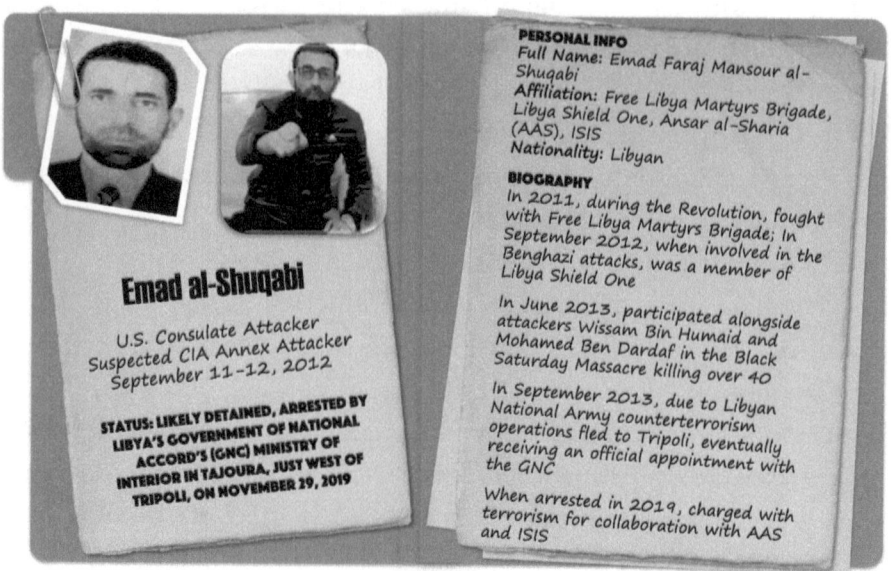

Emad al-Shuqabi

U.S. Consulate Attacker
Suspected CIA Annex Attacker
September 11-12, 2012

STATUS: LIKELY DETAINED, ARRESTED BY LIBYA'S GOVERNMENT OF NATIONAL ACCORD'S (GNC) MINISTRY OF INTERIOR IN TAJOURA, JUST WEST OF TRIPOLI, ON NOVEMBER 29, 2019

PERSONAL INFO
Full Name: Emad Faraj Mansour al-Shuqabi
Affiliation: Free Libya Martyrs Brigade, Libya Shield One, Ansar al-Sharia (AAS), ISIS
Nationality: Libyan

BIOGRAPHY
In 2011, during the Revolution, fought with Free Libya Martyrs Brigade; In September 2012, when involved in the Benghazi attacks, was a member of Libya Shield One

In June 2013, participated alongside attackers Wissam Bin Humaid and Mohamed Ben Dardaf in the Black Saturday Massacre killing over 40

In September 2013, due to Libyan National Army counterterrorism operations fled to Tripoli, eventually receiving an official appointment with the GNC

When arrested in 2019, charged with terrorism for collaboration with AAS and ISIS

Online, a former Libyan Minister of Health, made accusations that Wissam worked as an advisor in 2012 to the former Prime Minister Abdurrahim el-Keib. At the time, she reported that el-Keib had her meet with both Emad and Wissam's brother Qais bin Humaid—with Keib noting their importance as leaders in Libya. She believed that Keib was acting subservient to all three Wissam, Qais, and Emad and that they were siphoning off funds to treat wounded fighters. This is important as they likely then used those funds to support terrorist activities.

On September 11th, 2012, Emad attacked the U.S. Consulate in Benghazi, and was a suspect in the attack on the CIA Annex on September 12th, 2012. In June 2013, Emad participated in the Black Saturday massacre with Wissam and Benghazi attacker Mohamed ben Dardaf. He then relocated to Tripoli, in September 2013, fearing LNA counterterrorism operations. He was reported to have traveled between Tripoli and Misrata, as he was funneling weapons from both cities to then provide back to terrorists in Benghazi.

He then took on roles in the Libyan Government. First, he was appointed by the Libyan Ministry of Interior (MOI) in Tripoli to be the General Coordinator of the Intervention and Security Resolution Force starting in 2015. This position allowed him to have a diplomatic passport. Then he was given another appointment as an Adviser to the Libyan Investment Authority, one of Libya's largest sovereign funds. Since the end of the revolution, the Government of Libya had played a dangerous game allowing too many terrorists into the mix, making it hard to view it as a properly functioning or legitimate government.

In November 2019, he was arrested in Tajoura, just west of Tripoli. The arrest warrant noted that he was charged for being the coordinator between Ansar al-Sharia and ISIS.

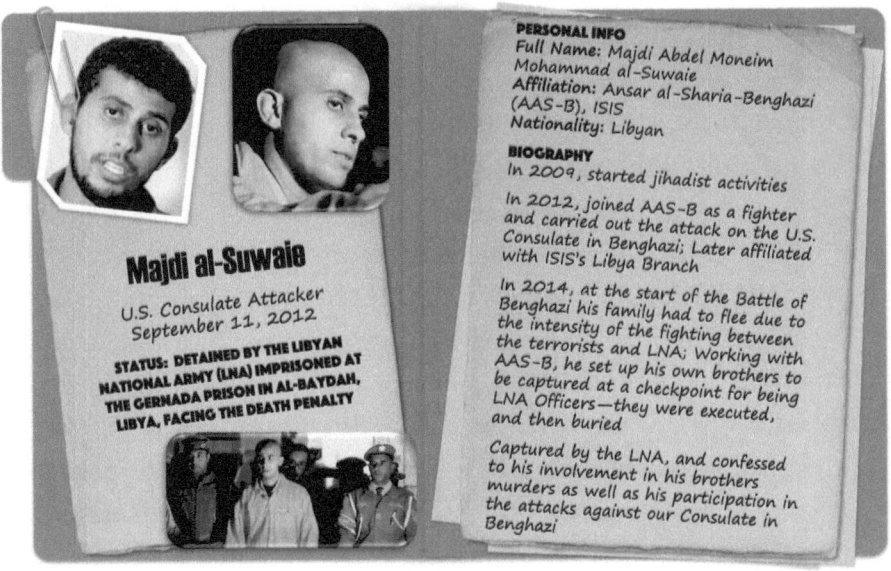

Majdi al-Suwaie
U.S. Consulate Attacker
September 11, 2012
STATUS: DETAINED BY THE LIBYAN NATIONAL ARMY (LNA) IMPRISONED AT THE GERNADA PRISON IN AL-BAYDAH, LIBYA, FACING THE DEATH PENALTY

PERSONAL INFO
Full Name: Majdi Abdel Moneim Mohammad al-Suwaie
Affiliation: Ansar al-Sharia-Benghazi (AAS-B), ISIS
Nationality: Libyan

BIOGRAPHY
In 2009, started jihadist activities

In 2012, joined AAS-B as a fighter and carried out the attack on the U.S. Consulate in Benghazi; Later affiliated with ISIS's Libya Branch

In 2014, at the start of the Battle of Benghazi his family had to flee due to the intensity of the fighting between the terrorists and LNA; Working with AAS-B, he set up his own brothers to be captured at a checkpoint for being LNA Officers—they were executed, and then buried

Captured by the LNA, and confessed to his involvement in his brothers murders as well as his participation in the attacks against our Consulate in Benghazi

(108) Majdi al-Suwaie, full name Majdi Abdel Moneim Mohammad al-Suwaie from the al-Laythi neighborhood in Benghazi. As is evident with the number of attackers from al-Laythi, it's the hotbed for extremists and terrorists. As noted previously the neighborhood was nicknamed "Kandahar, Libya"—an ode to the days' several militant groups like al-Qa'ida and LIFG were collocated in Kandahar, Afghanistan during the muj-era.

Majdi first became involved in terrorism in 2009. After the Libyan revolution, in 2012, he joined AAS-B. He became one of the group's key fighters, along with terrorists Mohammad Jaafar, Ayman al-Ashibi, Nasser al-Farsi, and Ahmed Khalifa. On September 11th, 2012, Majdi attacked the U.S. Consulate with fellow AAS-B member Mohammad Jaafar. In 2014, after the start of the LNA's Operation Dignity during the Battle of Benghazi, there was heavy fighting focused on the al-Laythi neighborhood between terrorists and the LNA. Families had to flee and become displaced persons in other parts of Libya. On December 24th, 2014, as Majdi's own family was departing Benghazi, he set up for his family to be stopped at an AAS-B checkpoint using his close associate Ahmed Khalifa from AAS-B, who went on to be a Senior Leader of ISIS in Libya. Majdi also supported ISIS in Benghazi.

During the stop, his three brothers, Ziyad, Adam, and Imad were kidnapped due to their roles serving as Military Officers in the LNA, with Ziyad serving as a policeman in Central Support, Adam serving as a soldier in Naval Forces, and Imad serving in the Military Police. The three were transferred to the Qawarsha area in Benghazi, where 17 February and AAS-B forces maintained a stronghold at the Qawarsha Gate from 2014 to 2016. At this location, Majdi's three brothers were executed and buried.

Majdi was captured by the LNA and confessed to his involvement in his brothers' murders and the September 11th Consulate attacks. Majdi was facing the death penalty for his crimes. As of 2022, he was imprisoned at the Gernada prison in al-Baydah, a maximum-security prison operated by the LNA holding high-value terrorists detained primarily in Benghazi and Darnah. We personally have zero interest in gaining access to or extraditing Majdi, as we look forward to the LNA putting him to death as soon as possible. The fate he most deserves as you do not kill your own brothers.

T

(109) Ali bin Taher, full name Ali Abdullah Muftah bin Taher with alias al-Far (Mouse). Before the Libyan revolution in 2011, Ali was accused of several criminal and terrorist charges, including theft and a terrorist plot to bomb a police academy and a police station.

After the Libyan revolution in 2011, Ali linked up with former Guantanamo Bay detainee Sufyan bin Qumo, a former member of al-Qa'ida in Afghanistan, and operated as his deputy in AAS-D. Ali was also linked to terrorists Hisham al-Falah, Zuhair al-Masli, and Abdel Salam bin Ali. On September 11, 2012, Ali traveled from Darnah to Benghazi to attack the U.S. Consulate.

Ali bin Taher

U.S. Consulate Attacker
September 11, 2012

STATUS: DECEASED, KILLED LIKELY BY AL-QA'IDA ON APRIL 7, 2014 ON A FARM OUTSIDE OF DARNAH, LIBYA

PERSONAL INFO
Full Name: Ali Abdullah Muftah bin Taher
AKA: Al-Far (Mouse)
Affiliation: Al-Qa'ida, Ansar al-Sharia-Darnah (AAS-D)
Nationality: Libyan

BIOGRAPHY
In 2011, was freed from Abu Salim prison at the start of the Revolution

At the time of the Benghazi attacks, was affiliated with al-Qa'ida and a member of AAS-D serving as a deputy to Sufiyan bin Qumo

In 2014, as a member of the Islamic State Army led a wave of the assassinations of policemen and judges in Darnah; On February 20, 2014, his attacks resulted in the closure of many polling stations in Darnah during the Constitution Drafting Assembly (CDA) vote

At the time of his death, he expected al-Qa'ida to assassinate him

At the end of the Consulate attacks, Ali stole an armored sedan used by our GRS team when it traveled to the Consulate. Ali returned home to Darnah after the attacks in our sedan. As a member of AAS-D, Ali was involved in killing many citizens in Darnah. He was also a threat to his own family after sending one brother into hiding, a failed attempt to kill an additional brother, shooting a nephew seven times, and forcibly marrying his sisters off to terrorists. Ali then moved on to an organization called the Islamic State Army. In March 2014, AAS-D and the Islamic State Army merged with Majlis Shura Shabab Al-Islam (MSSI) or Shura Council of Islamist Youth. Then the group declared Darnah an official Islamic Emirate.

On the day of his death on April 4th, 2014, Ali was paranoid that al-Qa'ida had put a hit out on him. He was killed that day at a farm outside of Darnah. Reportedly, Ali had stolen funds coming to al-Qa'ida, and specifically to Sufyan bin Qumo, from a joint operation with LIFG where they kidnapped the son of Libyan Special Forces Commander Colonel Wanis Bukhamada. Al-Qa'ida killed Ali as retribution for the theft; however, the murder affected Qumo so much that he broke ties with al-Qa'ida.

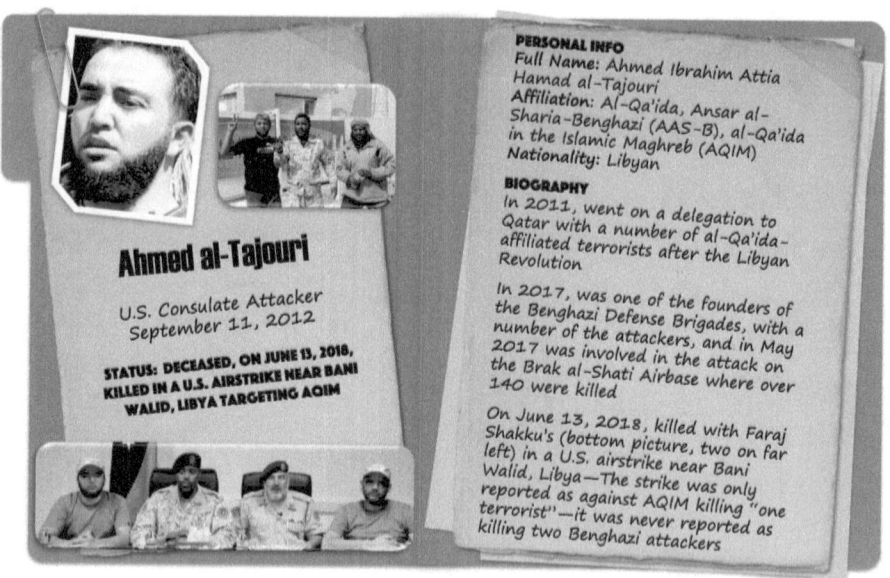

(110) Ahmed al-Tajouri, full name Ahmed Ibrahim Attia Hamad al-Tajouri from the al-Salmani neighborhood in Benghazi, was a member of al-Qa'ida. In 2011, he visited Qatar with a large group of terrorists affiliated with al-Qa'ida after the end of revolutionary fighting and the fall of the Gaddafi regime. On September 11, 2012, he attacked the U.S. Consulate in Benghazi with fellow Benghazi attacker Faraj Shakku.

In addition to al-Qa'ida, he became involved with AAS-B then BRSC after many Benghazi terrorist leaders died in battle during the Second Libyan Civil War. Key leaders killed included Wissam bin Hamid, Mohammad al-Zahawi, Yahya al-Maqsabi, Khaled al-Faydi, and Jalal Makhzoom. During the Battles of Benghazi, Tajouri established a covert militia coined the 17th Company with terrorists Yasser al-Jabali and Shakku. The 17th Company was based in the al-Sadadah area of Misrata under the protection of the May 28th militia.

The 17th Company came into existence as the LNA continued to defeat terrorists in Benghazi, more information was coming to light regarding just how much the Libyan Government was funding terrorists. Therefore, the Government felt that it was not advantageous to publicly partner with known militias who were actively involved in terrorism like those groups linked to the BRSC. Instead, the Libyan Government of Omar al-Hassi offered an alternative to Tajouri, noting that if he set up a militia, but named it like a security company, then the militia could still be contracted under the Ministry of Defense and continue to receive funding. This tactic took the pressure off the Libyan Government as it allowed for more of a hidden hand when financially supporting Islamists.

The former Leader of 17 February, Fawzi Bu Khatif also went on to form his own security company, coined Security Side. One location that his security company received a contract for was the Palm City Residences near Tripoli, Libya which was the primary expatriate community in the capital. He was also reported to have contracts with the Government of Turkey to train militias for the Libyan Government.

In 2017, Tajouri helped co-found the Benghazi Defense Brigades (BDB) with Ismail al-Sallabi, Faraj Shakku, and Yasser al-Jabali. Inside the city of Benghazi, he carried out many crimes and terrorist acts to include attacks on the army headquarters, courts, and security directorates.

On May 18th, 2017, he participated in the BDB attack on the LNA's Brak al-Shati Airbase, with 141 people killed, including 103 LNA soldiers. He appeared in online videos explaining his relationship with the leaders of al-Qa'ida. Before his death, he was a close associate of several AQIM-affiliated terrorists, including MBM, Abdel-Moneim al-Hasnawi, Ahmed Abd al-Jaleel al-Hasnawi, and Ibrahim al-Jadran.

Al-Qa'ida announced the deaths of Tajouri and Shakku in June 2018. Specifically, on June 13th, 2018, Tajouri and Shakku died in a U.S. airstrike near Bani Walid. As noted previously, the U.S. Government never reported the names of persons killed, noting that a strike against AQIM killed "one terrorist," so it again was not reported that two terrorists were killed who had participated in in the Benghazi attacks.

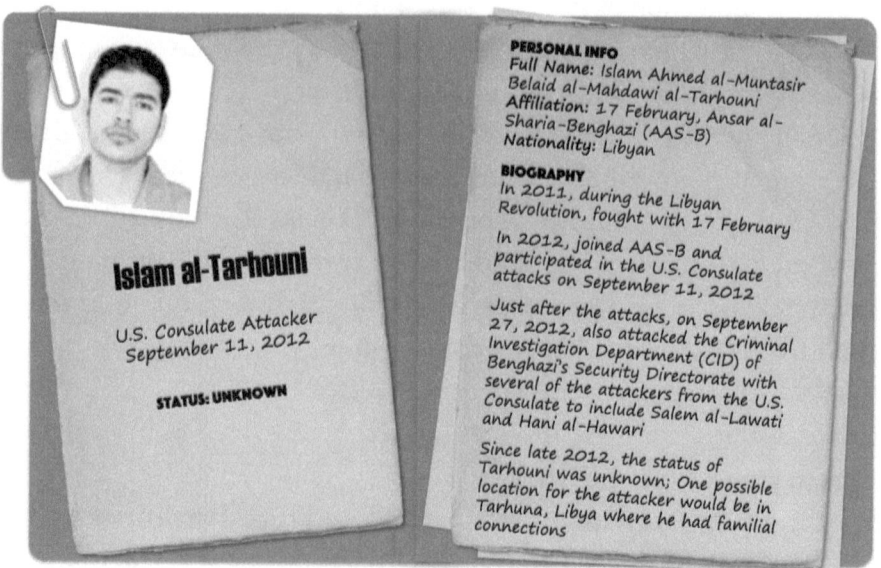

(111) Islam al-Tarhouni, full name Islam Ahmed al-Muntasir Belaid al-Mahdawi al-Tarhouni from Benghazi. Islam was arrested in 2007, charged with terrorism, and sent to Abu Salim prison. During the revolution, he fought with 17 February. In 2012, he joined AAS-B and participated in the attack on the U.S. Consulate in Benghazi on September 11th, 2012.

Just after the attacks, on September 27th, 2012, he also attacked the Criminal Investigation Department (CID) of Benghazi's Security Directorate with several of the attackers from the U.S. Consulate. Since 2012, the status of Tarhouni has been unknown. One possible location for the attacker would be in Tarhuna, Libya, where he had familial connections.

(112) Ramadan Trabelsi, with full name Ramadan Tawfiq Abdullah Trabelsi from Darnah. He was detained in 2007 and sent to Abu Salim prison. After likely being freed from prison during the revolution, there was only limited reporting available on him.

On September 11th, 2012, he participated in the attacks on the U.S. Consulate in Benghazi. It was unknown which group Ramadan was affiliated with at the time of the attacks, but it may have been the Abu Salim Martyrs Brigade (ASMB).

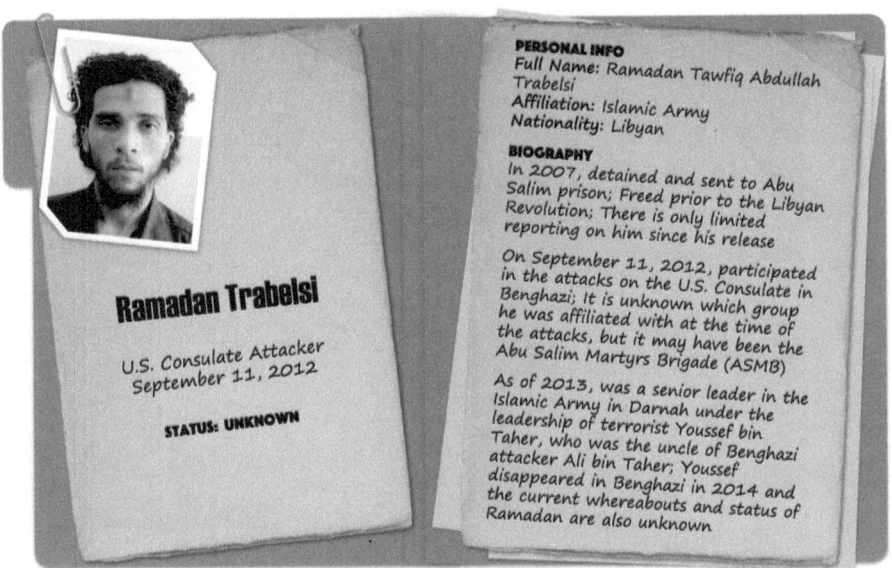

As of 2013, he was a senior leader in the Islamic Army in Darnah under the leadership of terrorist Youssef bin Taher, the uncle of Benghazi attacker Ali bin Taher. Within a year of the group being formed, Youssef disappeared under mysterious circumstances in Benghazi. In January 2015, Islamic Army, which was initially aligned with the DMSC, essentially switched sides in the terrorist battle for Darnah between al-Qa'ida and ISIS when it pledged allegiance to ISIS. Like Youssef, as of 2022, the status of Ramadan was also unknown.

U

(113) Faraj al-Chalabi, full name Faraj Husayn Hasan al-Shalabi al-Urfi, with alias Ahmad Abdallah al-Libi, was a core al-Qa'ida member who attacked the U.S. Consulate on September 11th, 2012. At the time of the attacks, Faraj was in contact with AQSL's Dr. Ayman al-Zawahiri, as well as with the leaders of al-Qa'ida's Yemen affiliate, AQAP.

As background, Chalabi was a previous LIFG member. He was named in a March 1998 arrest warrant with Osama bin Laden for the murder of two German tourists in Libya in 1994. In an interview, Chalabi said he left

Libya in June 1995, traveled to Sudan, and met up with Bin Laden before traveling to Syria. After leaving Syria, he traveled to the Pakistan and Afghanistan border region and settled in Jalalabad, Afghanistan, where he lived until the U.S. military operation in Afghanistan in October 2001. He then fled through the Tora Bora Mountains to Peshawar. Chalabi likely fled with Bin Laden or in proximity as he had previously served as a bodyguard to the al-Qa'ida Leader.

In 2004, Chalabi was arrested in a joint U.S. and Pakistani operation. At the time, he tried to escape, got up on a roof, and jumped off, but was injured and taken back into custody. Later the Pakistani Government extradited Chalabi to Libya, where he served in Abu Salim prison with several Benghazi attackers until being freed during the Arab Spring on February 15th, 2011.

On September 11th, 2012, Faraj attacked the U.S. Consulate in Benghazi. He took the attack one step further, collected documents from the Consulate, and then traveled back to Pakistan immediately after the attack to deliver the documents to al-Qa'ida Leader Zawahiri. There has been a lot of discussion regarding these documents, as they may have been sensitive and contained the personal information of Americans based in Benghazi. However, no classified holdings were stored at the U.S. Consulate in Benghazi. The only item in recent weeks that the CIA had even declassified to provide to the State Department was a listing of militia and terrorist camps in the region. So, any discussion that attackers, including Chalabi or maybe Abu Khatallah, as the FBI states, stole classified materials is patently false. Further, the CIA's Annex was never breached, and after the fact, they were able to collect and/or destroy any items still at the Annex in the month following the attacks.

At the behest of the U.S. Government, Chalabi was arrested in Pakistan in September 2012, then deported to Libya and arrested by Libyan security authorities at the airport in Libya's capital of Tripoli. He was taken to a military police prison. Chalabi was released on June 12th, 2013, due to a "lack of evidence," meaning the FBI did not release any or enough evidence to Libyan authorities to hold him for his crimes against our friends. Honestly, it's unclear if the U.S. even put in an official extradition request for Chalabi from Libya. On July 14th, 2014, Chalabi was captured by a local militia in Marj, Libya, for personal issues between him and the militia. He was tortured and found dead two days later. Jihadi life has consequences.

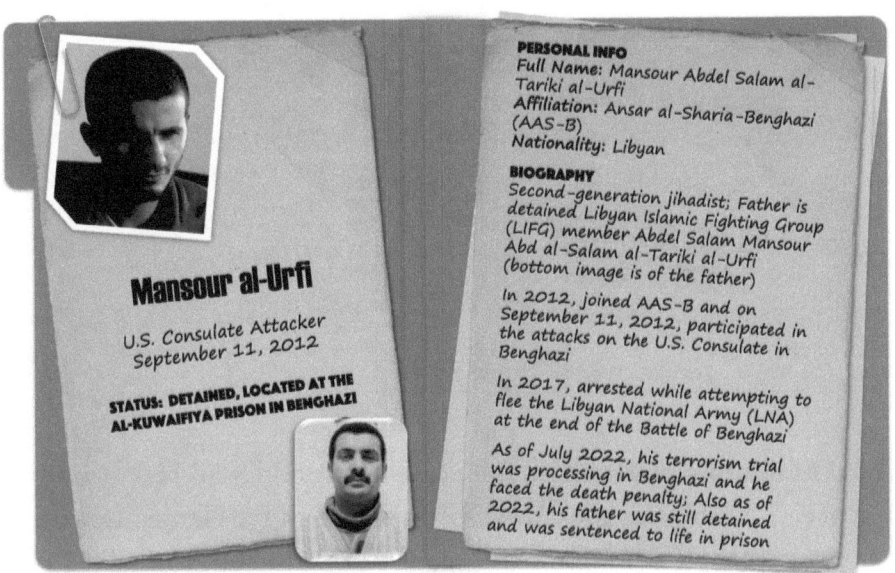

(114) Mansour al-Urfi, full name Mansour Abdel Salam al-Tariki al-Urfi from Benghazi. Before the Libyan revolution, Mansour was not known to have been involved in terrorism. However, Mansour is a second-generation jihadist. His father was Abdel Salam Mansour Abd al-Salam al-Tariki al-Urfi, a LIFG member. Abdel was sentenced to prison after his arrest in 1996 for preparing to carry out a near-term attack with several terrorists from Palestine, Syria, and Libya.

In 2012, Mansour joined AAS-B and, on September 11th, 2012, he participated in the attacks on the U.S. Consulate in Benghazi. In 2017, he was arrested while attempting to flee the LNA at the end of the Battle of Benghazi. Mansour was detained at the al-Kuwaifiya prison in Benghazi. As of July 2022, his terrorism trial was in process, and he faced the death penalty. His father was also still detained.

Mahmoud al-Wahishi
U.S. Consulate Attacker
September 11, 2012
STATUS: DECEASED, WAS KILLED IN SEPTEMBER 2014 BY THE LIBYAN NATIONAL ARMY (LNA) DURING THE BATTLE OF BENGHAZI

PERSONAL INFO
Full Name: Mahmoud Saad Hussein Abdul Ghani al-Wahishi
AKA: Abu Dujana
Affiliation: Al-Qa'ida in the Islamic Maghreb (AQIM)
Nationality: Libyan

BIOGRAPHY
In approximately 2003, left Libya to join al-Qa'ida to fight in Iraq; Stopped in Algeria and joined AQIM instead

In 2011, returned from Algeria to fight in the Libyan Revolution; Along with fellow Benghazi attacker Anis al-Houti and other AQIM members, founded the Uthman bin Affan Brigade and its Al-Ghuraba Brigade which included a number of foreign fighters

In 2012, attacked the U.S. Consulate with his brother Hussein; Starting in 2014 fought with al-Qa'ida terrorists against the LNA during the start of the Second Libyan Civil War and attempted to overthrow LNA's Saiqa camp

(115) Mahmoud al-Wahishi, full name Mahmoud Saad Hussein Abdul Ghani al-Wahishi with alias Abu Dujana was from the Garden District in Benghazi. Mahmoud was a member of AQIM. In approximately 2003, he left Libya to travel to Iraq to join al-Qa'ida to fight U.S. forces. On the route, he stopped in Algeria, joined AQIM, and passed on follow-on travel to Iraq—never joining AQI.

In 2011, Mahmoud returned from Algeria to fight in the Libyan revolution. He founded the Uthman bin Affan Brigade along with fellow Benghazi attacker Anis al-Houti and other AQIM members. They also set up a unit in the Brigade called the Al-Ghuraba Brigade, which included a number of foreign fighters, most of whom were from Tunisia.

On September 11th, 2012, he and his brother, Hussein al-Wahishi, were both attackers at the U.S. Consulate in Benghazi. Over the next few years, Mahmoud was consistently involved in disputes with various terrorist organizations as they jockeyed for priority in Libya. He also fought with al-Qa'ida terrorists against the LNA during the Battle of Benghazi. Mahmoud was involved in attempts to overrun LNA camps in Benghazi and was killed by the LNA in September 2014. His fellow attacker brother, Hussein, as of 2022, was at-large.

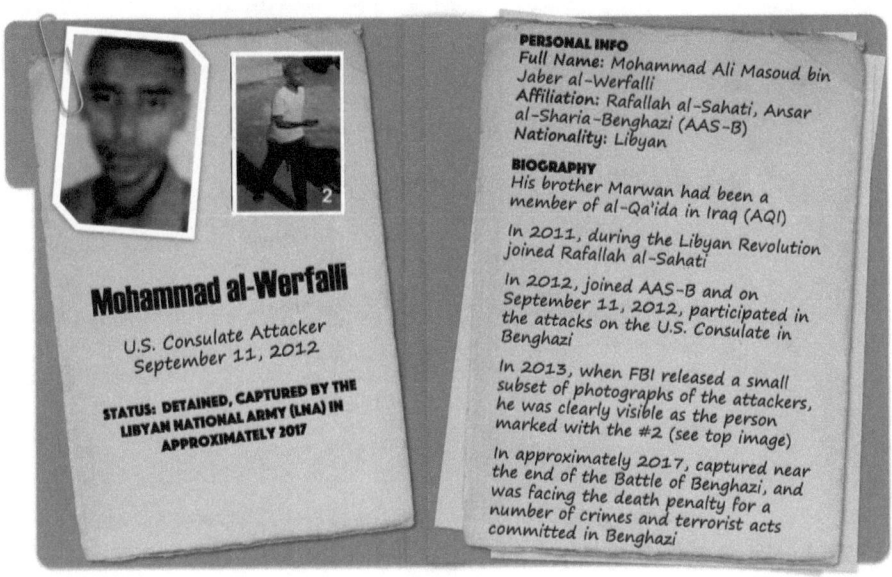

(116) Mohammad al-Werfalli, full name Mohammad Ali Masoud bin Jaber al-Werfalli from Benghazi. Mohammad was not known to have been involved in terrorism before the Arab Spring. However, his brother Marwan al-Werfalli, full name Marwan Ali Masoud bin Jaber al-Werfalli was a member of AQI. Marwan left Libya in the mid-2000s to fight in Iraq, and his status post the Iraq War was unknown.

In 2011, during the Libyan revolution, Mohammad joined Rafallah al-Sahati Brigade, and in 2012, he joined AAS-B. On September 11th, 2012, Mohammad participated in the attacks on the U.S. Consulate in Benghazi. In 2013, when the FBI released a small subset of photographs of the attackers, Mohammad was visible in the lineup.

In approximately 2017, near the end of the Battle of Benghazi, the LNA captured Mohammad. He is facing the death penalty for several crimes and terrorist acts he committed in Benghazi. As of 2022, he was located at the al-Kuwaifiya prison in Benghazi.

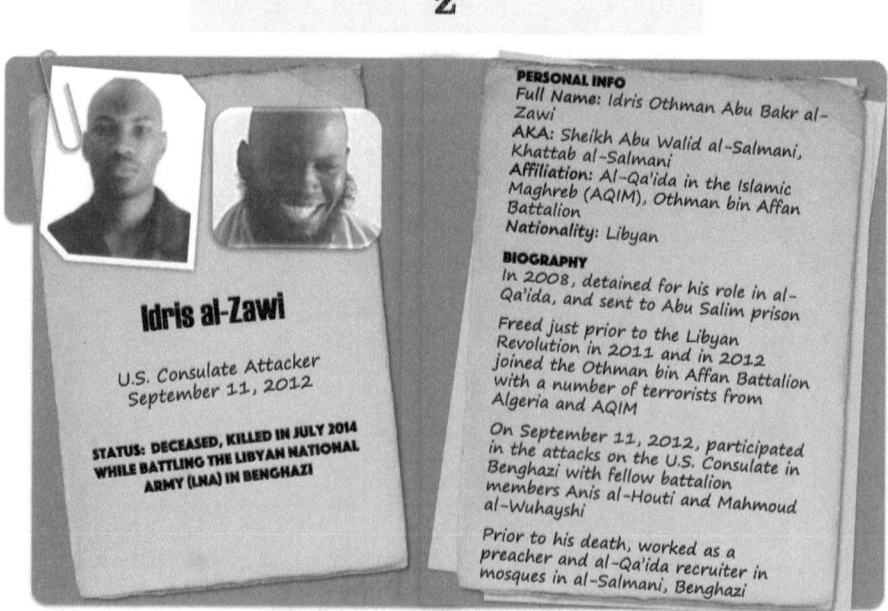

(117) **Idris al-Zawi**, full name Idris Othman Abu Bakr al-Zawi with aliases is Sheikh Abu Walid al-Salmani and Khattab al-Salmani from Benghazi. Idris was a member of AQIM. He was arrested in 2008 for his role in al-Qa'ida and sent to Abu Salim prison. He was released just before the Libyan revolution in 2011 and, during the revolution, was an outspoken religious preacher who promoted violence. In 2012, he joined

the Othman bin Affan Battalion with several terrorists from Algeria that he had fought with in AQIM.

On September 11th, 2012, he participated in the attacks on the U.S. Consulate in Benghazi with fellow Othman bin Affan battalion members Anis al-Houti, and Mahmoud al-Wahishi. Throughout the next year, terrorists took almost complete control of the city of Benghazi due to the security vacuum. Idris took advantage of this situation and used his preaching skills in the mosques of the al-Salmani neighborhood in Benghazi, where he successfully recruited large numbers of terrorists who were then sent to conflict areas.

In May 2014, Idris joined in with terrorists to battle the LNA in Benghazi, and he was killed in July 2014 by the LNA.

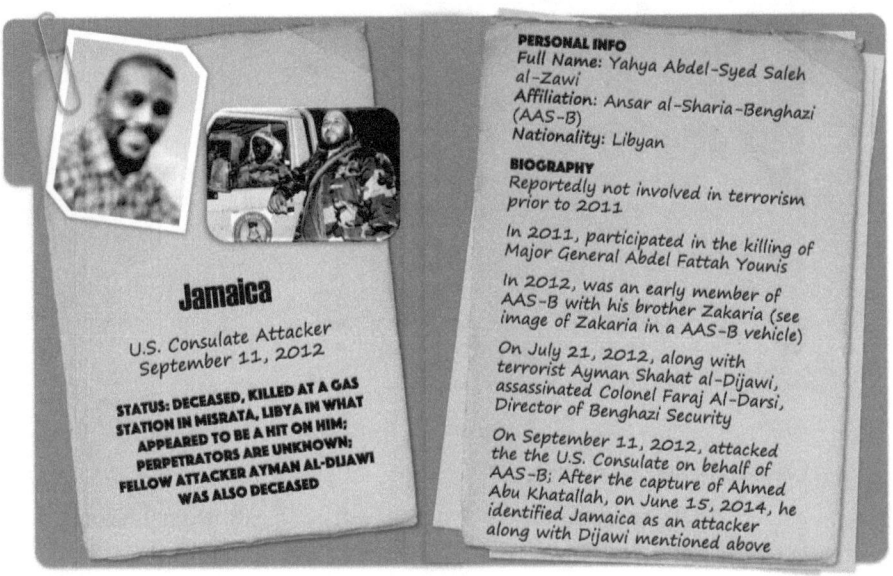

(118) **Jamaica**, real name Yahya Abdel-Syed Saleh al-Zawi from the al-Sabri area of Benghazi. Before the Libyan revolution in 2011, Jamaica was not involved in terrorism. On July 28th, 2011, he participated in the most high-profile assassination in the city of Benghazi, which was the killing of Army Commander Major General Abdul Fatah Younis al-Obeidi. Then approximately one year later, on July 21st, 2012, he participated in the assassination of Colonel Faraj al-Darsi, Director of Benghazi Security,

with associates Ayman Qarun and Ayman Shahat al-Dijawi. During the revolution, Dijawi had fought with the Rafallah al-Sahati Brigade.

In 2012, Jamaica joined AAS-B. On September 11th, 2012, Jamaica and Dijawi, participated in the U.S. Consulate attacks. Jamaica's brother, Zakaria, was also a member of AAS-B. In April 2013, Jamaica was killed at a gas station in Sirte, accompanied by a group of foreigners from the Gulf. The murder of Jamaica appeared to be a hit, but it was unknown who carried out the murder. Fellow attacker Dijawi was also deceased.

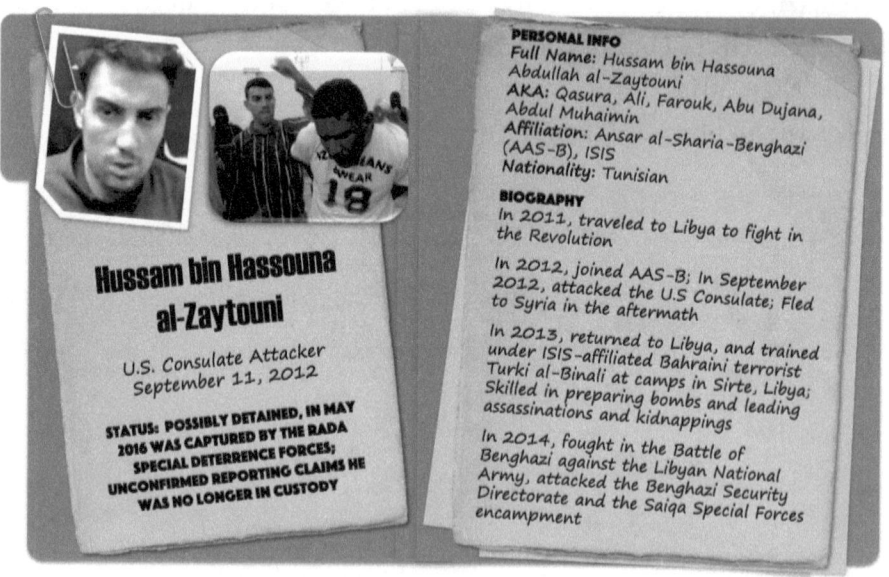

Hussam bin Hassouna al-Zaytouni

U.S. Consulate Attacker
September 11, 2012

STATUS: POSSIBLY DETAINED, IN MAY 2016 WAS CAPTURED BY THE RADA SPECIAL DETERRENCE FORCES; UNCONFIRMED REPORTING CLAIMS HE WAS NO LONGER IN CUSTODY

PERSONAL INFO
Full Name: Hussam bin Hassouna Abdullah al-Zaytouni
AKA: Qasura, Ali, Farouk, Abu Dujana, Abdul Muhaimin
Affiliation: Ansar al-Sharia-Benghazi (AAS-B), ISIS
Nationality: Tunisian

BIOGRAPHY
In 2011, traveled to Libya to fight in the Revolution

In 2012, joined AAS-B; In September 2012, attacked the U.S Consulate; Fled to Syria in the aftermath

In 2013, returned to Libya, and trained under ISIS-affiliated Bahraini terrorist Turki al-Binali at camps in Sirte, Libya; Skilled in preparing bombs and leading assassinations and kidnappings

In 2014, fought in the Battle of Benghazi against the Libyan National Army, attacked the Benghazi Security Directorate and the Saiqa Special Forces encampment

(119) Hussam bin Hassouna al-Zaytouni, full name Hussam bin Hassouna Abdullah al-Zaytouni with aliases Qasura, Ali, Farouk, Abu Dujana, and Abdul Muhaimin from Tunisia. In 2011, Zaytouni traveled to Libya to fight in the revolution. In 2012, he joined AAS-B. On September 11th, 2012, he participated in the attacks on the U.S. Consulate in Benghazi. In the aftermath of the attacks, he fled to Syria, fearing U.S. reprisals. In 2013, Zaytouni returned to Libya and trained under Bahraini terrorist and ISIS-affiliate Turki al-Binali at camps in Sirte, Libya. Zaytouni was skilled in preparing SVBIEDs. He also led assassinations and kidnapping operations.

In 2014, Zaytouni participated in the Battle for Benghazi and joined the BRSC. Zaytouni was fighting with terrorists when they overran the Saiqa Special Forces encampment. He also attacked the Benghazi Security Directorate and was on video, hitting a detainee who was later found dead. In May 2016, he was likely captured by the RADA Special Deterrence Forces in Tripoli. However, unconfirmed reporting claims he was no longer in custody as of 2022.

CHAPTER 8

Attack Network Analysis Part I

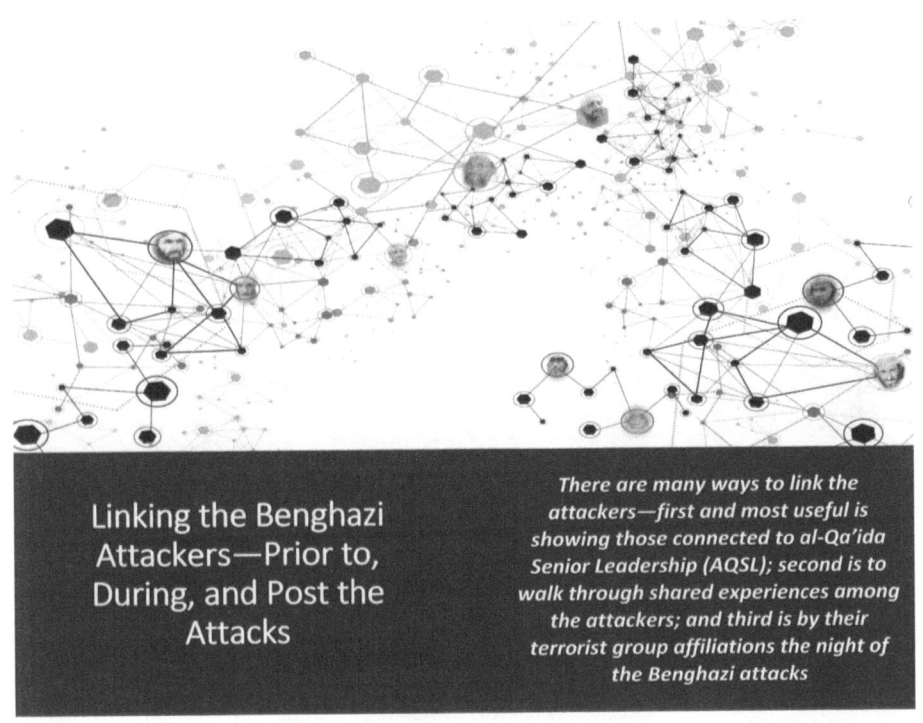

Linking the Benghazi Attackers—Prior to, During, and Post the Attacks

There are many ways to link the attackers—first and most useful is showing those connected to al-Qa'ida Senior Leadership (AQSL); second is to walk through shared experiences among the attackers; and third is by their terrorist group affiliations the night of the Benghazi attacks

As the Benghazi attacks culminate decades of terrorist relationships and collective experiences, the following analysis section will display the importance of those links. The first key link was those attackers with historical connections to al-Qa'ida Senior Leadership (AQSL), particularly to the last two now-deceased leaders of al-Qa'ida, Osama bin Laden, and Dr. Ayman al-Zawahiri. A second key connecter among attackers, besides the majority being from eastern Libya, was the significant number of attackers who had served in Abu Salim prison together—many with lengthy sentences for terrorism. The last and most common way to show the relationship among attackers is through terrorist group memberships.

Former al-Qa'ida Leader Dr. Ayman al-Zawahiri (Zawahiri hereafter) trusted this network of militants to attack us, the Americans in Libya. Part of this trust came from serving together in Afghanistan. As background, the Soviet-Afghan War started in 1979, but it was not until the mid-1990s, at the behest of Osama Bin Laden, that the bulk of the "Afghan Arabs" traveled to the region in support of the Afghan resistance fighters that would later become the Taliban. It is no secret that we at the CIA were heavily involved in Afghanistan at the time as it was a crucial node in the Cold War where the CIA believed it could affect Soviet influence in the region. As such, the CIA, along with others like the Brits, the Pakistanis, and Saudis, helped fund what would become the Afghan mujahideen and would fund the early days of Bin Laden's al-Qa'ida organization. Obviously, like with our support to the Libya revolution, aligning with Islamists usually has dangerous repercussions for the U.S., its allies, and its interests.

What is surprising when you look back is that Bin Laden used to frequent our American Club at the U.S. Consulate in Peshawar. This area was known as the wild wild west, called Pakistan's North West Frontier Province, a suitable name that's now been changed to Khyber Pakhtunkhwa province. Bin Laden would sit and drink tea along with U.S. diplomats and CIA officers. Ironically, U.S diplomats and CIA officers later filtered through Peshawar to attempt to track and locate Bin Laden and his associated terrorist allies in the region.

In addition, the U.S. Consulate attack in Benghazi was directed by Zawahiri from AQSL's base in Pakistan. Just two years prior, AQSL aligned with local Pakistani Taliban allies attacked the U.S. Consulate in Peshawar on April 5th, 2010. At the time, al-Qa'ida primarily used two groups in Pakistan to pull attackers from being the Tehrik-e-Taliban Pakistan (TTP) and Lashkar-e-Jhangvi. The attackers arrived at the Consulate with a vehicle-borne improvised explosive device (VBIED) and on foot to assault the compound. Luckily the VBIED got stuck on a barricade a local guard activated. No Americans were killed, but the attack killed eight locals, including trusted guards at the compound.

A quick sidebar, as is well-known, just before the attacks on September 11th, 2001, Osama Bin Laden had the foresight to know that the U.S. would respond in kind, and that would mean American boots on the ground in Afghanistan. The CIA had already been trying to align with a new regional ally, Ahmad Shah Massoud, the then leader of the Northern Alliance, who controlled the northeastern portions of Afghanistan. Massoud, an ethnic Tajik, was a local warlord more aligned with American ideals as he fought for a pro-democratic Afghanistan and renounced the Taliban.

Massoud was likely humored at the new position as CIA's golden boy as the U.S. Government had been trying to persuade him to surrender to the Taliban for half a decade. Massoud stayed steadfast that there was no benefit in siding with Islamic extremists--we still don't heed his warning. If only the U.S. had also understood the repercussions of what happens when you join forces with Islamists to help you win wars, maybe there would have been a first 9/11, but maybe not a second 9/11 incident in Libya in 2012.

Bin Laden knew he had to remove Massoud. What is not as commonly known is that he trusted one person to provide the two Tunisian suicide bombers for the attack. That person was Seifallah Ben Omar Ben Hassine, with alias Abu Ayyad al-Tunisi. As noted previously, Hassine was the Leader of Ansar al-Sharia-Tunisia (AAT) at the time of the September

Attack Network Analysis

2012 Benghazi attacks. Hassine personally met with both Bin Laden and Zawahiri to plan the assassination of Massoud.

Massoud was assassinated just two days before 9/11. Hassine also provided terrorists to the attacks against us in Benghazi on September 11, 2012, and he was personally on the ground to lead the storming of the U.S. Embassy in Tunis, Tunisia, just days later on September 14th, 2012. Al-Qa'ida was still using the same people and networks during 2012 that they used in 2001. So, if you do not get that September 11th, 2001, and September 11th, 2012, are not intimately connected, then you are not paying attention.

The following lays out leaders and attackers with direct links to the attacks in Benghazi who had served in the Pakistan and Afghanistan border region with al-Qa'ida.

In addition to ties to AQSL, all of al-Qa'ida's global affiliates played a role in the attacks. The three we would like to highlight are AQIM, the historical network of AQI, and AQE, which was a newly formed group in 2012. As Zawahiri directed AQIM to plan the attacks, the group had the most significant number of attackers within al-Qa'ida. The Benghazi attacker ranks included over 35 terrorists from AQIM who were reported to be on the ground at the U.S. Consulate.

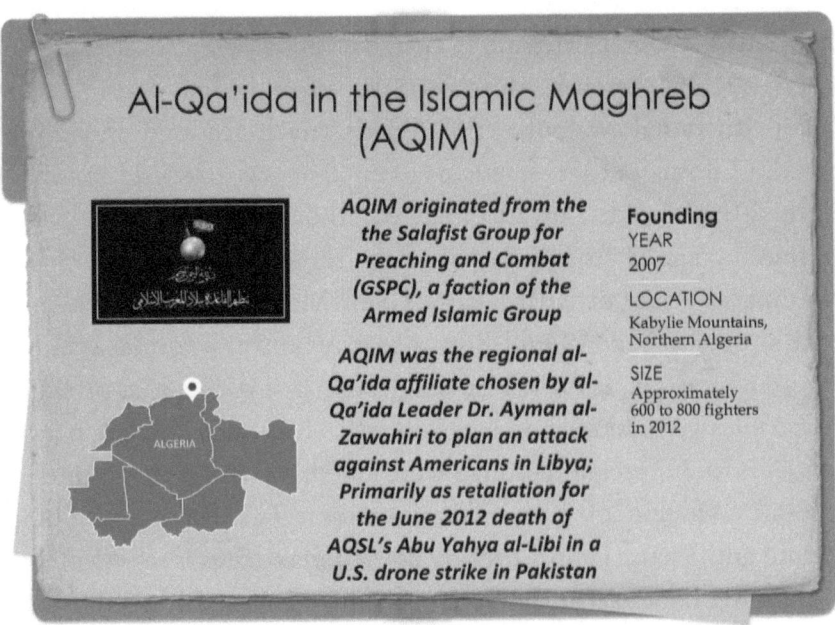

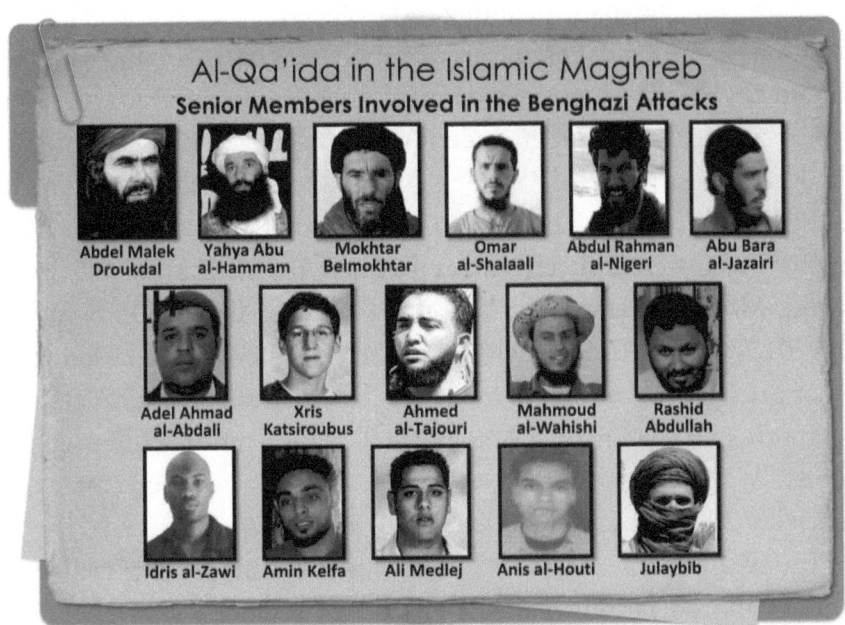

AQIM also used several different battalions across the group's Saharan Brigades for the attack. The most public facing battalion being the al-Mulathameen Battalion led by AQIM's MBM. This battalion not only carried out the U.S. Consulate in Benghazi attacks, it also carried out the January 2013 In Amenas, Algeria attack.

For situational awareness, this Algeria attack occurred only 25 miles from the Libyan border; it was assessed that the attackers trained and launched from inside Libya to carry out the attack. The attack was also reportedly planned from Libya after the target was recommended by a long-time associate of MBM named Sidi Mohamed Ould Bouamama, better known as Sanda Ould Bouamama, who was a former spokesman for Ansar Dine, an al-Qa'ida-linked group. In the event, approximately 800 people were taken hostage, and over 130 were foreign nationals. Sadly, at least 39 foreign nationals were killed, including three Americans: Frederick Buttacio of Katy, Texas; Gordon Lee Rowan of Sumpter, Oregon; and Victor Lynn Lovelady of Nederland, Texas.

Both the Benghazi and Algeria attacks were masterminded by MBM, working with AQIM's Yahya Abu al-Hammam and AAS-B's Mohammad al-Zahawi for the plot planning. The Algeria attack also shared an operational commander on the ground with Benghazi, Abu Bara al-Jazairi. The attacks shared 11 more individual attackers located on the ground at both events. This figure included Abu Bara's fellow Benghazi attackers Xris Katsiroubus, and Ali Medlej. As noted previously, both Xris and Ali died in the attacks. The other nine were lower-level Egyptian AQIM members who were only referred to by their aliases: Sahib, Mohammed, Towfiq, Mohsin, Issa, Salem, Yousef, Fahad, and Battar. On September 9th, 2013, AQIM issued a video on the In Amenas attacks eulogizing two of the Egyptian attackers to include listed Battar as "Battar al-Masri" and another attacker named Abu Hajar al-Masri. In sum, counting inside sources, AQIM used just over 30 attackers during the Algeria attack, and reportedly 29 of those were killed during the siege.

The two attacks also shared several persons supporting facilitation, training, and operations, particularly AQIM senior facilitator Al Hasan Ould Khalil, best known as Julaybib, who served in a similar role to a Chief of Staff in AQIM. He handled much of the communications between MBM and AQIM's leader Droukdal. He was also the son-in-law of MBM. In addition, AQIM trainer and operational leader Abdul Rahman al-Nigeri, with aliases Abu Dujanna and Abdassamad supported both the Benghazi and Algeria attacks, but was only confirmed to be located on the ground during the Algerian attack. He died in Algeria, as well.

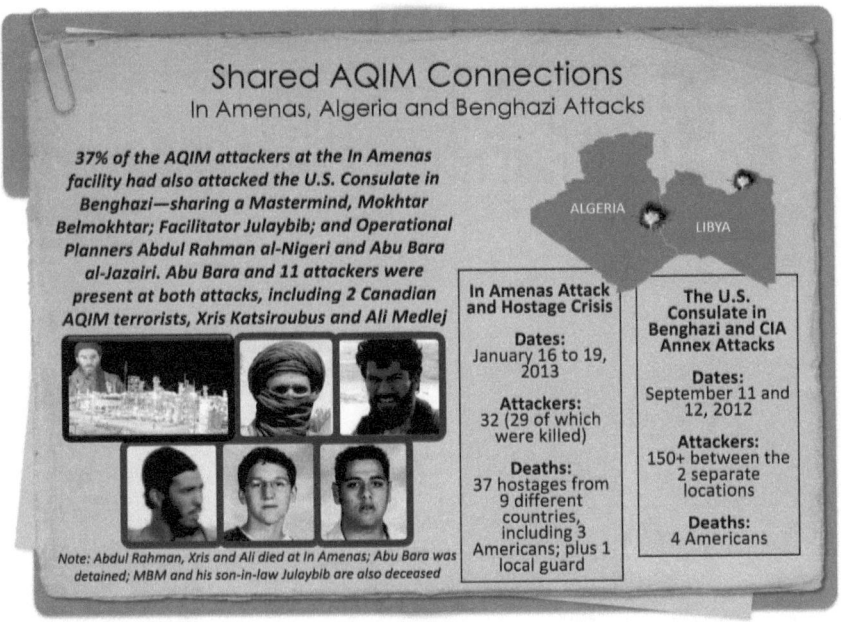

Besides the al-Mulathameen Battalion, we would like to shine a spotlight on the involvement of an additional key AQIM battalion referred to as the Mali Group. We highlight this group, as while there is limited reporting on the group and its activities, they provided at least a dozen attackers for the U.S. Consulate in Benghazi attacks. The Libyan Mali Group members included Mahdi Dango, Mansour al-Shalaali, Jafaar Azzouz, Abu Hamza al-Tabawi, Khaled al-Ammari, Taher al-Awami, Abdul Hamid al-Shaeri, Hassan al-Shukri, and Hashem Bousidra; and the Egyptian Mali group members included Asmi Ahmed, Ahmed Hazaa, Karim Moawad al-Rahmani, and Mohammad Jaber Abdel Wahab.

Attack Network Analysis

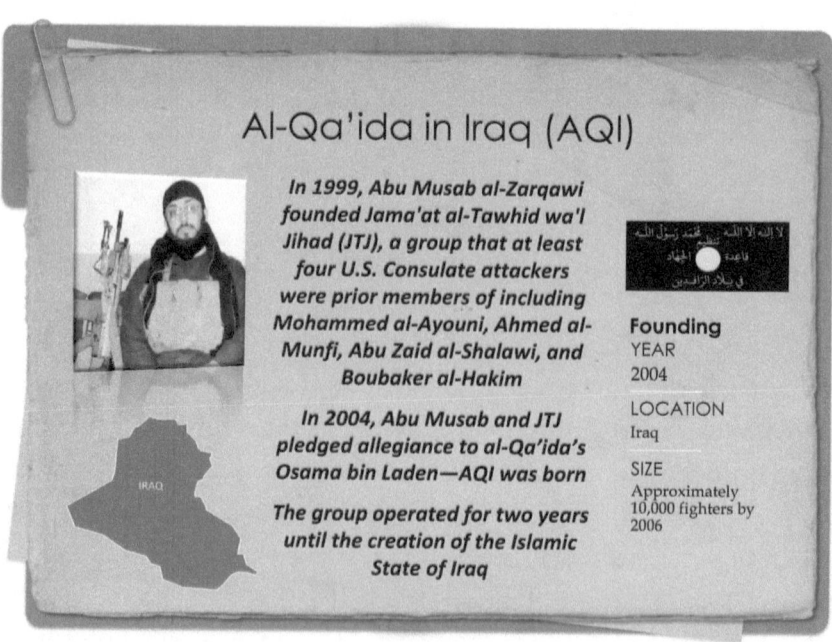

The insurgency in Iraq in the 2000s relied heavily on foreign fighters, and Libya was one of those countries that provided willing recruits in droves. As such, a number of the Benghazi attackers either attempted to join AQI, did join AQI, fought in Iraq, and/or had family members who died while fighting U.S. forces in Iraq.

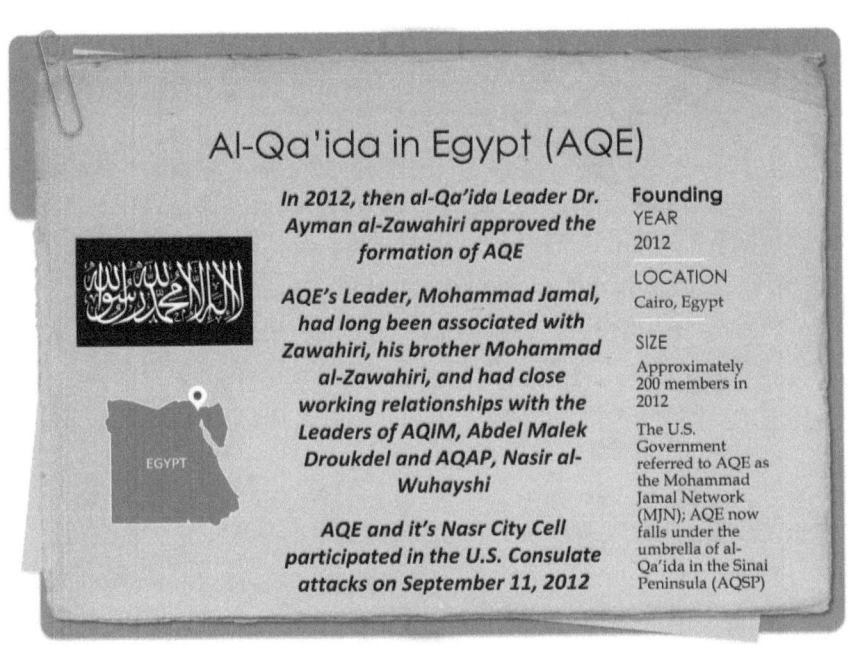

Another regional affiliate of al-Qa'ida that provided attackers to support the Benghazi attacks was AQE, which is a part of an umbrella organization that now falls under al-Qa'ida in the Sinai Peninsula (AQSP). Former EIJ member Mohammad Jamal led the group. If you have followed Benghazi over the years, U.S. intelligence officials have referred to these attackers as the Mohammad Jamal Network, likely a bit of a ploy to distance the group from al-Qa'ida. But the fact is al-Qa'ida Leader at the time, Zawahiri, had received and approved an official request before the attacks from Jamal to form al-Qa'ida in Egypt. The group was being funded by AQAP's Leader Nasir Abdel Karim al-Wuhayshi (hereafter Wuhayshi), who, not long after the attacks, in 2013, Zawahiri appointed Wuhayshi as his deputy in AQSL.

Jamal formed AQE after his release from an Egyptian prison in 2011. Like with groups in Tunisia and Libya, he took advantage of the significant release of terrorists in Egypt by the new government post the Arab Spring uprisings. Due to his direct involvement in the Benghazi attacks, Jamal was re-arrested by Egyptian authorities in November 2012. His computer was exploited, and there was documentation to include letters to Zawahiri in which Jamal asked for assistance to acquire weapons, conduct terrorist training, and establish additional terrorist groups in the Sinai. Jamal's key terrorist groups formed before his arrest were AQE and an operational cell coined the Nasr City Cell under AQE.

Egyptian authorities reported nine other individuals were arrested who may have had a direct role in the September 11th, 2012, Benghazi attacks and who were believed to be members of the Nasr City Cell. Four of the named Benghazi attackers included:

1. Karim Ahmed Essam el-Azizi variant Karim Ahmed Essam al Azizy with alias Karim al-Badawi, from Egypt. He was killed during an October 24th, 2012 raid on the Nasr City Cell.
2. Nabil Abdel-Moneim from Egypt.
3. Sami Maghraby with alias Abu Basir from Egypt.
4. Mohammed Saeed Merghany from Tunisia.

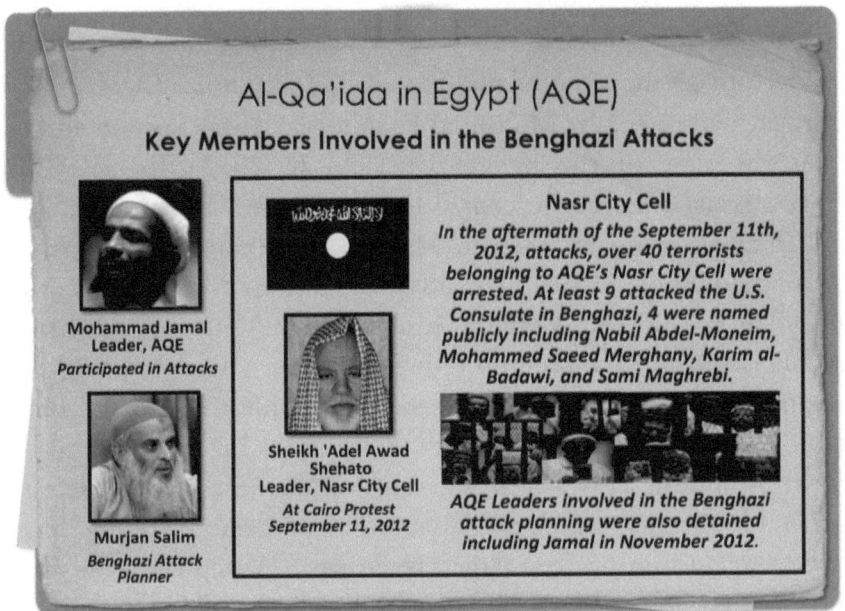

As an aside, to take a quick step back before the attacks, in 2011, after being released from prison, Jamal spared no time re-establishing his old connections from Afghanistan, Yemen, and from within Egypt. He was in direct contact with AQIM Leader Droukdal, and, as noted AQAP's Leader Wuhayshi, who was killed in a U.S. drone strike in Yemen on June 12th, 2015. He was also a contact with Wuhayshi's Deputy, Qasim Yahya Mahdi al-Rimi, better known as Qasim al-Rimi. Rimi, led AQAP after Wuhayshi's death. Rimi was a thorn in our side at the CIA for a very long time. In addition, one of our close friends lost his brother, William "Ryan" Owens, on January 29th, 2017, in a U.S. Special Operations raid that had just missed capturing Rimi in Yemen. Rimi finally met his demise on January 29th, 2020, due to a U.S. airstrike in the Yakla area of Wald Rabi' District, Al Bayda Governorate, Yemen.

One of the issues that made Rimi such a lasting and formidable foe to us at the CIA was that he always remained focused on al-Qa'ida External Operations and had previously led a plot targeting the U.S. Ambassador in Sanaa, Yemen. Part of his focus on the West and primarily the U.S. was because his brother, Ali Yahya Mahdi al-Rimi, had long been held in Guantanamo Bay, Cuba, arriving on May 2th, 2002, and not released until April 16th, 2016. Another key al-Qa'ida commander that Jamal was close to at the time related to al-Qa'ida External Operations was Khorasan Group Leader Muhsin al-Fadhli, with alias Dawood al-Assady.

At the time of the Benghazi attacks in 2012, this relationship between Jamal and Fadhli seemed less concerning, as little was known about Khorasan's goals. Then in May 2013, three al-Qa'ida members, Mohammed Abdel-Halim Hemaida Saleh, Amir Mohammad Abu-al-Ila Aqidah variant Amr Mohammed Abu al-Ela Aqida, and Mohammad Mustafa Mohammad Ibrahim Bayyumi were arrested in Egypt for planning attacks on foreign Embassies to include the U.S. Embassy in Cairo. Detainee Saleh noted that he "believed in conducting attacks against American and Israeli interests" and confessed to "recruiting suicide bombers to send to Syria and had been planning terrorist activities against unspecified targets in Europe." As shown in the graphic, the attackers reported that both Jamal and Fadhli were directing their activities. Jamal was hiding from authorities after the Benghazi attacks, so his connection to this group was likely before September 2012—which means AQSL was potentially planning attacks on Americans in Libya and Egypt simultaneously and using some of the same attackers.

Fadhli's Khorasan Group was relatively unknown until September 2014, when the U.S. conducted strikes in Syria that were reported to be against the group as it posed an "imminent threat" at the time to U.S. interests—meaning the group was in the final planning stages of a homeland attack. What was known, though, is that Fadhli was directed by and reported solely to Zawahiri, showing Zawahiri's direct role and a keen interest in carrying out attacks against the West. Like the CIA's concerns with Rimi, Fadhli always viewed al-Qa'ida External Operations as the most critical mission for al-Qa'ida.

Attack Network Analysis Part II

Links Between the Benghazi Attacks and Instigators of the Protest at the U.S. Embassy in Cairo

The Protests in Cairo, Egypt during the early evening hours of September 11th, 2012 did not occur in a vacuum, as key individuals involved in inciting the protests, to include Mohammad al-Zawahiri, were directly involved in al-Qa'ida's planned attack against the U.S. Consulate in Benghazi later that evening

While the Egypt protests did not roll over into neighboring eastern Libya on September 11th, 2012, it is still pertinent to our story. One key item that is not always explained is that several individuals who were aware of the planning and plot to attack the U.S. Consulate in Benghazi were also involved in inciting the violence in Cairo. The two involved in both the Benghazi attacks and the Cairo protests are Mohammad al-Zawahiri, again the brother of former al-Qa'ida Leader Dr. Ayman al-Zawahiri, and Sheikh Adel Shehato, the leader of the primary operational cell for AQE, the Nasr City cell. Before the Arab Spring, Zawahiri and Shehato served in prison with fellow AQE members linked to the Benghazi attacks, including Mohammad Jamal, Tarik Abu al-Azzam, Murjan Salim, and Karim al-Badawi. As they knew the attacks were occurring later that night in Benghazi, we believe they were incentivized to further promote the protests in Cairo to shift U.S. resources towards Cairo, leaving Benghazi more vulnerable. We found out after the fact that the U.S. military, even though they knew of the protest in Cairo over ten days in advance, did not posture any resources to prepare, nor did they scramble any DoD resources to respond once the protests were occurring in Cairo.

As background, the protest in Cairo was planned to start in late August 2012 by al-Gama'at Al-Islamiyah (IG) variant Jamaat al-Islamiyya. Yes, planned before any nonsense regarding a video was known; again, another reason we are not discussing the fake video narrative. Specifically, on August 30th, 2012, the group officially announced the September 11th, 2012 protest in Cairo. The supporters were asked to protest the ongoing incarceration of the terrorist group's leader Sheikh Omar Abdel-Rahman better known as "The Blind Sheikh". Mohammad al-Zawahiri was one of the main leaders of the protest, along with the Blind Sheikh's family and former senior EIJ leaders like Shehato, Sheikh Tawfiq al-Afani, Rifai Ahmed Taha Musa, and Ahmed Ashoush. Note that Ashoush was the founder of Ansar al-Sharia in Egypt. The IG protestors then took advantage of the video narrative and used it to promote the video after the fact. This included Ashoush issuing a fatwa on September 16th, 2012, calling for the makers of the "Innocence of Muslims" film to be killed.

As the protests were planned in honor of the Blind Sheikh, besides the Benghazi-related personalities being involved like Mohammad al-Zawahiri and Shehato, the Blind Sheikh's son Mohammed Abdel-Rahman with alias Asadullah had been involved in several protests involving the release of his father in Egypt throughout 2012. Asadullah was a former CIA prisoner held in a black site like former LIFG leader Khalid al-Sharif. As background, in the early 1980s, Asadullah joined the fight in Afghanistan like many of our Benghazi attackers. He was captured by U.S. allied forces in 2003. After time in CIA custody, he was returned to Egypt, where he spent several years in jail with many of the AQE attackers in our story. He was released in the fall of 2010.

Back to his father, in 2012, the Blind Sheikh was serving a life sentence in the U.S. for involvement in the 1993 World Trade Center bombing. The Blind Sheikh was one of the main targets al-Qa'ida was attempting to kidnap Ambassador Stevens in exchange for. Other attacks in Benghazi were in the Blind Sheikh's honor, including the attack on the British Ambassador's convoy on June 11th, 2012. Throughout 2012, there was a big regional push by militants and pro-Muslim Brotherhood politicians as they truly believed the Blind Sheikh could be released as part of the successes of the Arab Spring. In June 2012, then Egyptian President-elect Mohamed Morsi vowed publicly to free the Blind Sheikh during his inaugural address. Even in the months preceding the attacks through January 2013, Morsi was still working towards releasing the Blind Sheikh. At the time, this wasn't a pipe dream; there was plenty of buzz in the U.S. Government that the Obama Administration was considering transferring the Blind Sheikh to Egyptian custody.

In addition to Blind Sheikh's freedom being one of several motivators for the Benghazi attacks, which included the death of Abu Yahya al-Libi, the Blind Sheikh was also part of negotiations during the in Amenas oil facility siege in 2013. Separately, Ahmed Omar Abdul Rahman, another of the Blind Sheikh's sons, was killed in a U.S. airstrike in Afghanistan on October 15th, 2011. EIJ issued a eulogy at the time. Just a month after the Benghazi attacks, in October 2012, Zawahiri put out a statement on Ahmed's death and vowed again to get his father released and ask for kidnappings against Americans to help seal the deal as Ambassador Steven's kidnapping did not come to fruition.

One last anecdote back to AQSL, on September 10th, 2012, Zawahiri released his video eulogizing his Deputy Abu Yahya al-Libi. In the 42-minute video, Zawahiri urged Muslims, particularly Libyans, to avenge Abu Yahya: "His blood urges you and incites you to fight and kill the crusaders," he said. Zawahiri requested his death be avenged in Libya, just one day before the Benghazi attacks. He directed the attacks and was well aware the attacks were forthcoming.

Attack Network Analysis Part III

Attack Associates Who Served in the Abu Salim Prison

In 2008 and 2010 terrorists started to be released from the Abu Salim Prison in Tripoli due to international pressure, primarily from the U.S. Government who removed Libya from its list of State Sponsors of Terrorism

Those released, by 2011, had become the strongest base of the opposition forces to Libya's Leader Moammar Gaddafi

While much of the historical relationships within AQSL stemmed from fighting in Afghanistan, those Libyans, who remained with core al-Qa'ida well through the U.S. invasion of Afghanistan in 2001, became senior leaders in the group. These senior leaders included Abu Yahya al-Libi, Attiyah Abd al-Rahman, and Abu Layth al-Libi—all former members of the LIFG who were killed in U.S. drone attacks in Pakistan. Abu Yahya was killed in June 2012, Attiyah in August 2011, and Abu Layth in January 2008.

For Libyans who met a different fate and were captured in the Pakistan and Afghanistan border region, they were returned to the Gaddafi regime ended up in the notorious maximum security Abu Salim prison. The prison was named after the neighborhood in Tripoli, where it was located. Even those deported back to Libya after first serving in Guantanamo Bay, Cuba were also sent the Abu Salim prison. For most of these terrorists once they entered the prison, it was believed they might spend their lives there if they did not die an early death unnaturally due to mistreatment. In a horrific example, on June 29th, 1996, a massacre in the prison killed over 1200 inmates.

On May 15th, 2006, then Secretary of State Condoleezza Rice announced that the United States was removing Libya from its list of state sponsors of terrorism. Soon after, the U.S. resumed diplomatic relations. With this change in Libya's designation, Saif al-Islam Gaddafi, the son of Dictator Gaddafi, partnered with international envoys and held reconciliation talks with LIFG and other organizations representing the detainees. The primary mediator of the talks was prisoner Ali al-Sallabi, again the brother of Ismail al-Sallabi, who founded the 17 February and its more radical offshoot, the Rafallah al-Sahati Brigade. Talks were successful for the Islamists and led to approximately 90 LIFG detainees being released in April 2008 and 233 additional detainees being released in March 2010. By September 2010, according to Libyan officials, more than 700 prisoners who were detained for being Islamists had been released.

By 2011, these former detainees became the strongest base of the opposition to Gaddafi. Those extremists left behind in the prison who were not released during reconciliation were freed during the uprising in 2011. The alliances that formed among the Abu Salim detainees was formidable, their relationships were based on survival and trust, leading to a warrior class that, unfortunately, translated to a reality where it became terrorists first in Libya and revolutionaries last.

As the majority of attackers had a direct or familial links to Abu Salim prison and/or the groups established to honor those who served in the prison including the Martyrs Group and the Abu Salim Martyrs Brigade, we want to instead highlight one specific case from 2007. This select Libyan counterterrorism case from the Gaddafi era, had at least

18 terrorists named in it that went on to be U.S. Consulate in Benghazi attackers in September 2012. The 18 had been captured and detained by the prior regime for supporting global terrorist networks, including al-Qa'ida.

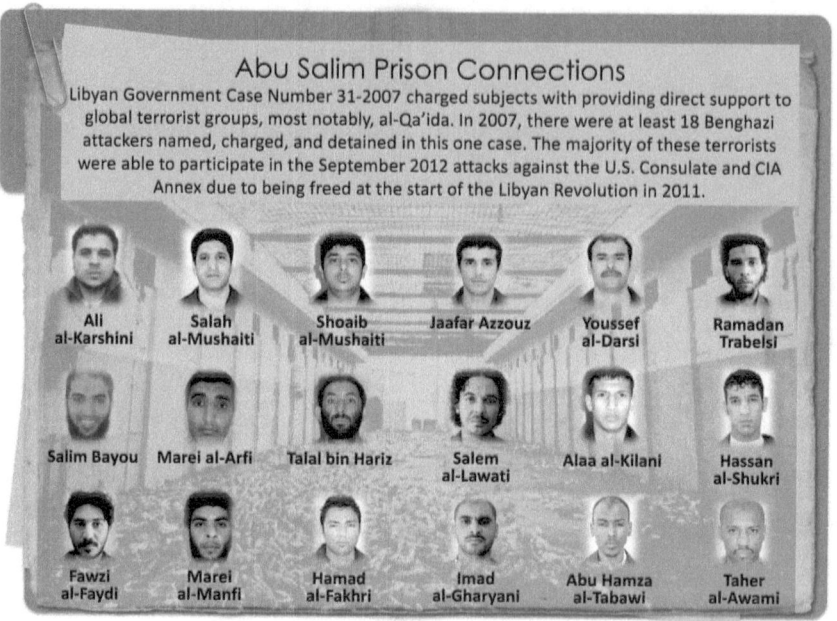

Attack Network Analysis Part IV

This book mentions five individual groups using the moniker "Ansar al-Sharia" connected to the Benghazi attacks network. We will set Ansar al-Sharia-Sirte and Ansar al-Sharia-Egypt aside and note there was also an Ansar al-Sharia in Yemen. The three to flag most pertinent to the Benghazi attacks included Ansar al-Sharia-Benghazi (AAS-B) led by Mohammad al-Zahawi; Ansar al-Sharia-Darnah (AAS-D) led by Sufyan bin Qumo with Benghazi attacker Ali bin Taher as his right-hand; and Ansar al-Sharia-Tunisia (AAS-T) led by Seifallah Ben Hassine. As AAS-B was al-Qa'ida's primary attack ally in Benghazi for the U.S. Consulate attacks on September 11th, 2012, we will delve into that group specifically.

Al-Qa'ida's Key Benghazi Attack Ally, Ansar al-Sharia Benghazi (AAS-B)

In February 2012, as al-Qa'ida was months into establishing a base in Benghazi, it's key local affiliate, AAS-B formed with the goal to establish an Islamic caliphate; The groups initially worked seamlessly with some members being dual-hatted as AAS-B was an umbrella organization; AAS-B also expanded the initial Benghazi assassination cells created in 2011 by al-Qa'ida's Senior Libyan Operative Abu Anas al-Libi

In February 2012, Zahawi was one of the founders of AAS-B. He founded the terrorist umbrella organization with a number of Benghazi-based extremist allies, including Nasser al-Tarshani, Ali al-Karshini, Mahmoud al-Barassi, Fawzi al-Faydi, Mansour al-Faydi, Mohammad Awad Trabelsi, Al-Hussein al-Sheikhi, Jafar Mohammad Omar Bayou, Bashir Suwaid, Ayman Shahat al-Dijawi, Al-Mu'tasim Billah al-Nayhum, Ibrahim al-Torjoman, Nassib Fannoush, Mohammad Terbol, Mohammad al-Falah, Fawzi Khatal al-Subhi, Fawzi Butbel, Ahmed Mertah, Marwan al-Fazani, Moussa al-Karami, Hussain al-Gamal-Maddouri, Houjeen al-Mahdawi, Ahmed bin Nasser Karim, Bras Ali, Fayez Attiyah, Mahmoud and Yousef al-Awami. Most founders were killed in the Battle of Benghazi against the LNA, while dozens of other lower-level AAS-B terrorists managed to escape to western Libya.

By the summer of 2012, AAS-B grew exponentially to close to 1,000 members pulled from various militia organizations and terrorist groups operating in the region. After their first annual conference in June 2012, which brought in militants from all of North Africa and was a show of force similar to military parades of nations like Russia and North Korea, AAS-B was the lead-up and coming Islamist organization influencing Benghazi life. While the group first focused on humanitarian efforts in Benghazi to win the hearts and minds of locals, they shifted quickly by realizing their strength would come from attacks, not charity.

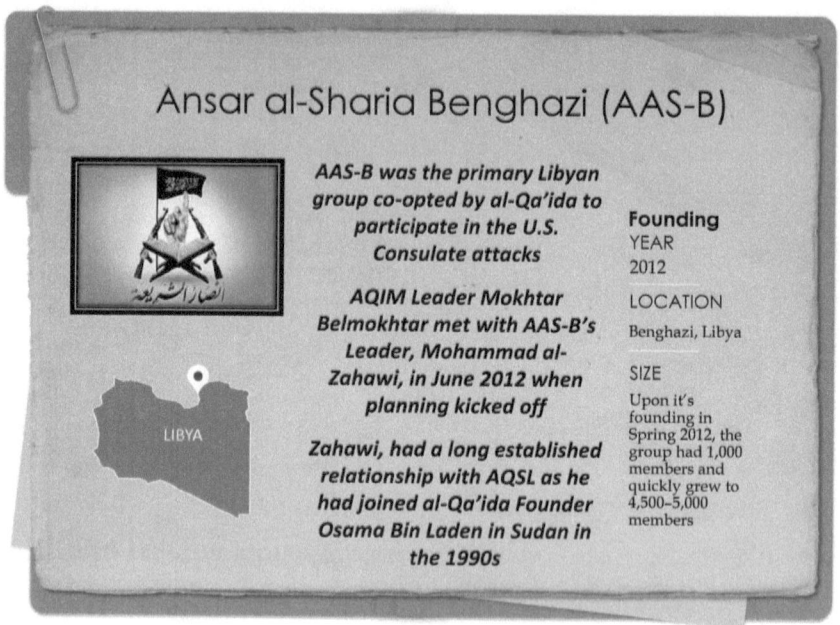

On June 18th, 2012, a group of 20 armed AAS-B members stormed the Tunisian Consulate in Benghazi and raised their black flag over the Consulate. This attack was almost three months before the assault on our compounds. Only 20 attackers overran the Consulate, while 7 to 8 times the number of attackers assaulted the U.S. Consulate and CIA Annex in September 2012. While Department of State security personnel in Benghazi requested and tried to improve security measures at the Consulate, no improvements were made as a result of the attack on the Tunisians.

As the number of U.S. Consulate attackers associated with AAS-B was significant, and as it was an umbrella organization, so many terrorists were members of AAS-B and then another regional terrorist group, we pulled out a select number to highlight in the included graphic.

Ansar al-Sharia Benghazi
Select Members Involved in the Benghazi Attacks

L-R: Mohammad al-Zahawi, Abu Khalid al-Madani, Mahmoud al-Awami, Faraj Shakku

L-R: Ahmed bin Nasser Karim, Talal bin Hariz, Fawzi al-Faydi, Hamza el-Sallak and Ahmed al-Mushaiti

L-R: Mahmoud al-Barassi, Hassan al-Karami, Mohammad al-Obeidi, and Abdullah Bouzkia

L-R: Khaled al-Saqzli, Ramadan al-Rubaie, Imad al-Gharyani, and Ahmed al-Majbari

Al-Qa'ida and its affiliates had many terrorist allies who supported the Benghazi attacks in September 2012. In the Brothers in Arms section, we walked through all the attackers, their group memberships, and historical affiliations. The al-Qa'ida attack allies that had been the most public were AAS-B, AAS-D, Rafallah al-Sahati Brigade, and 17 February Martyrs Brigade. Therefore, we are taking a moment to highlight a group that not only stayed under the radar but was believed to be defunct when they have continued to remain active. That was the Martyrs Group.

As noted in the biography of Benghazi attacker Mohammad al-Kawil, who was assessed as of August 2022 to be the current Leader of the group, the Martyrs Group was established in the aftermath of the Abu Salim Prison massacre in June 1996. The group's first act in the 1990s was to free injured terrorists from the al-Jalaa Hospital in Benghazi.

The group was based in Benghazi, Libya, and operated a cell in Europe. In June 2022, Switzerland expelled a long-time member of the Martyrs Group, Salah Ramadan al-Fitouri. Switzerland gave the terrorist asylum in 1998 and provided him funds through social assistance programs. Salah had directly supported terrorism in Libya, Syria, and Iraq.

On September 11th, 2012, at least nine Benghazi attackers were

Attack Network Analysis

affiliated with the Martyrs group. One of those members, Ayman Bouamoud, was close friends with Abdel Wahab Qaid, brother of Abu Yahya al-Libi. Again, the U.S. Consulate attacks were carried out partly to honor Abu Yahya after his death.

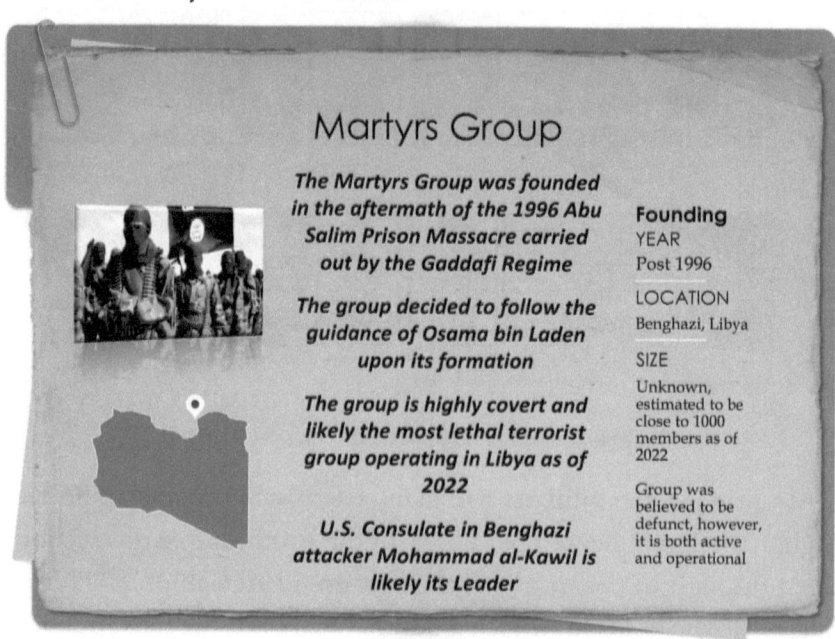

Attack Network Analysis Part V

Al-Qa'ida's Additional Local Attack Allies

Above are Mohammad al-Gharabi, al-Qa'ida's senior Commander in Benghazi in 2012; Ismail al-Sallabi, co-founder of 17 February/Rafallah al-Sahati Brigade and al-Qa'ida's deputy commander Benghazi; Salem Darby, Leader, Abu Salim Martyrs Brigade; Jalal Makhzoum, Benghazi attacker and co-founder, al-Qa'ida's al-Farouq Brigade; Khaled Mertah, brother of Benghazi attacker Ahmed al-Munfi; and Ashraf bin Ismail, key terrorist financer for our Benghazi attack network

While al-Qa'ida and its allies affected the operation at the U.S. Consulate in Benghazi, a local ally then also took advantage of the circumstances that evening to carry out follow-on attacks on the CIA Annex. Al-Qa'ida ended their attacks before the end of the day on September 11th, 2012; however, our CIA Annex started to be attacked in the early hours of September 12th, 2012. There were no indications that al-Qa'ida traveled over to the attacks, so it was assessed that one local group, which ended up being the Libya Shield, carried out all the attacks on the CIA Annex that day. These attacks led to a lethal mortar strike directed by terrorist Mastermind, Wissam bin Humaid. At the time, Wissam was the Leader of Libya Shield One. He had also co-opted Boka al-Oraibi, the Leader of Libya Shield Two, into assisting him in setting up the perfect circumstances where all Americans would be co-located so they could affect a direct strikes on us.

Attack Network Analysis

Unlike the Consulate, reporting related to the CIA Annex attacks is virtually non-existent. We only investigated terrorists who held senior positions within Libya Shield in September 2012 or those Libya Shield members who were already confirmed within our investigation to have participated in the attacks on the U.S. Consulate in Benghazi. For accuracy, all the CIA Annex attackers are labeled as suspected.

Attack Network Analysis Part VI

Attackers Who Joined the Islamic State

In the years following the Benghazi attacks, a number of attackers moved on from local militias and historic al-Qa'ida affiliations to join the more extreme Islamic State (ISIS), which founded a branch in Libya in November 2014

Following the Benghazi attacks, a tranche of attackers from local militias and those with historic terrorist affiliations within EIJ and al-Qa'ida joined the Islamic State of Iraq and the Levant. The group was also known as the Islamic State of Iraq and Syria and the Islamic State (known hereafter as ISIS). ISIS was increasingly more extreme than al-Qa'ida and the Libyan militias who fought against Gaddafi. Obviously, as evident in this book's pages, the attackers at the U.S. Consulate in Benghazi were not a one-off grouping of first-time offenders. Our attackers were hardened jihadists, fighting for decades before 2012 and planning to fight for as long as they could survive after.

ISIS in Libya is a whole story in and of itself. So to keep it brief, in 2014, ISIS officially established a base in Sirte, Libya. In its first couple of years, ISIS had momentum and swiftly gained 10,000 members by 2017. They had the membership numbers to rival any holdover militia group during the revolution. Not every Islamist who joined or swore allegiance to ISIS stayed with ISIS, especially as militants kept moving west towards the capital of Tripoli to fight the LNA. In Tripoli, terrorists tended to join local militias as ISIS had no power base in the capital and as militias were well-funded by the Government of Libya.

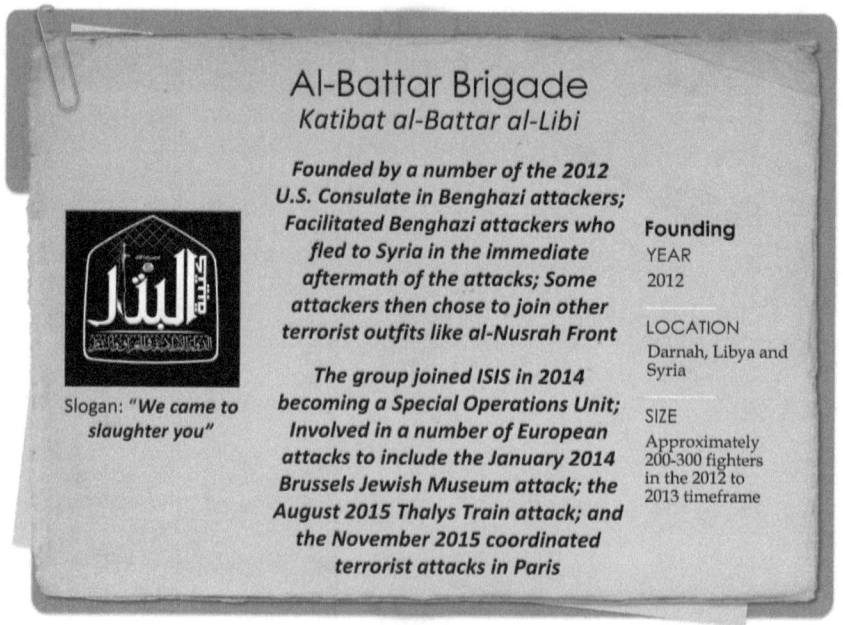

Al-Battar Brigade
Katibat al-Battar al-Libi

Founded by a number of the 2012 U.S. Consulate in Benghazi attackers; Facilitated Benghazi attackers who fled to Syria in the immediate aftermath of the attacks; Some attackers then chose to join other terrorist outfits like al-Nusrah Front

The group joined ISIS in 2014 becoming a Special Operations Unit; Involved in a number of European attacks to include the January 2014 Brussels Jewish Museum attack; the August 2015 Thalys Train attack; and the November 2015 coordinated terrorist attacks in Paris

Slogan: "We came to slaughter you"

Founding YEAR
2012

LOCATION
Darnah, Libya and Syria

SIZE
Approximately 200-300 fighters in the 2012 to 2013 timeframe

ISIS and al-Qa'ida were aligned at first as they both viewed General Haftar (and the LNA) as a shared enemy when he kicked off counterterrorism operations termed Operation Dignity in 2014—first in Benghazi and then in Darnah. The two terrorist groups could not stay aligned, and war broke out between them for primacy in several locations, most notably in Darnah. At this point, some attackers returned to their roots with al-Qa'ida and fought for the group. So, as many attackers joined ISIS and then flip-flopped back to al-Qa'ida, we will highlight a subset of the attackers who created a group in Libya in 2012 called the al-Battar Brigade, which became a key ISIS operational cell.

In 2012, Benghazi attacker Mahdi Saad al-Ghaythi co-founded the al-Battar Brigade Syria Branch, with the name variant Katibat al-Battar al-Libi. The Brigade originally consisted of defectors from al-Qa'ida-affiliate Abu Salim Martyrs Brigade (ASMB) and other extremist groups operating in Benghazi, Ajdabiya, and Tripoli. Mahdi was a key Military Official and member of its Security Committee. He helped facilitate fellow attackers from Benghazi to Syria as they fled Libya in the aftermath of the attacks fearing U.S. retaliation for the killing of an Ambassador.

By 2014, the al-Battar Brigade pledged allegiance to ISIS and essentially became the Libyan wing of ISIS. Members affiliated with this Brigade were later involved in supporting the November 13th and 14th, 2015, coordinated terrorist attacks in Paris, France—killing 130; and the May 22nd, 2015 attack at the Ariana Grande concert in Manchester, United Kingdom—killing 22.

Chapter 9

Our Final Thoughts

Losing an Ambassador in and of itself is usually the worst outcome you can imagine when serving in a diplomatic posting overseas, as it is both a professional and a deeply personal loss. The loss of the optimism of Chris Stevens, the mind of Sean Smith, Rone's leadership, and Bub's spirit makes this so much harder to bear. All four dedicated their lives to serving their country, with over 80 years of service combined. Their time in service does not seem to align with the failure of justice afforded to their murderers by the U.S. Government. Think back to the attackers we mentioned in this book. Like us, these men have trained together and bled together in a host of foreign wars. These attackers had allegiances and alliances dating back farther than any of us in Benghazi had even known each other.

When you look at these attackers, how many were known to you before this book for being involved in the Benghazi attacks? How many did you see on a wanted list produced by the FBI or the State Department's Rewards for Justice Program? Even as they killed two of State's own. One of the biggest questions should be, why were you fed a Mastermind who didn't even know of the attacks until after they started? Did that deliver any true justice, or was it so the FBI could get a conviction and allow the U.S. Government another fake narrative regarding Benghazi. Try having served in Benghazi to hear half the U.S. population not only call it a "conspiracy" but make jokes about it or infer that your friends deserved to be killed.

Look at these attackers, the dates some met their demise, and the fact that justice came through a Libyan Military Commander that the U.S. and NATO do not support. We will not get into current local politics in Libya, but it is essential to understand that our country backs a current Government in Libya that does support terrorists. We need to hold the U.S. Government responsible for making choices that align with American ideals ahead of the elites and European ambitions for the

Our Final Thoughts

country of Libya, all of which frankly probably evolve around taking advantage of the widespread corruption within the Libyan Government. In August of 2022, they couldn't even keep the lights on for their citizens in Benghazi.

These events caused complete paralysis across the U.S. Government as Benghazi was misused as a political football where no one wanted the ball, and no one wanted to be blamed for dropping the ball. Benghazi, though, caused issues across other warzones that those outside our community rarely are able to see. After Benghazi, every small, isolated base with little support was paranoid that they would be pulled out and shut down if threats were misconstrued.

This hesitancy in reporting resulted in incidents being downplayed, reported incorrectly purposefully, or not reported at all. Even during the FBI's trip to Benghazi, I asked a CIA Security Officer what DoD resources would be positioned closer to Benghazi to provide potential crisis response. He noted that he and the COS in Tripoli decided not to ask for DoD support from AFRICOM. The reasoning...they did not want it to seem the city was too dangerous and could thus bar us from returning with an Annex in the future. So not only did we trust Wissam bin Humaid with the security protection of our CIA colleagues and visiting FBI investigators in October 2012, we had no contingency if they came under attack, were kidnapped, or worse on the visit to Benghazi.

After the Benghazi attacks, Boon's first redeployment in October 2012 was back to a new warzone, in the Near East. He witnessed firsthand some Benghazi aftermath nonsense when the CIA facility was probed by suspected al-Qa'ida members conducting surveillance. They used the ruse of posing as potential home buyers and attempted to solicit information from our local guard force. This activity was observed on CCTV, and two CIA Officers went to investigate. This was the third al-Qa'ida surveillance team that had recently visited the compound. The suspected surveillants responded in broken English that they just wanted to buy our property, despite repeatedly being told it was not for sale. During the entire incident, our local guard force kept telling CIA staff that the hopeful buyers were, in fact, al-Qa'ida and that this was a threat.

Several days later, Boon's Security staff went to the CIA Case Officer, who engaged with the local guards about the potential attackers. The Case Officer wrote up the incident. However, his COB chose not to submit the information to CIA operational channels because of fears that CIA Headquarters would consider evacuating the Base to prevent another Benghazi. Several weeks later, the Base received additional threat reporting that al-Qa'ida was operationally targeting a house where Americans lived and described our house down to the color and exact layout. Yet, the COB and other CIA Staff also dismissed that reporting. They claimed the threat was against a separate house, when not only were we the only Americans living anywhere in the vicinity, but there was also no nearby house that was the same color. Security became an afterthought, as CIA Officers wanted to get their warzone credit and were not taught the leadership failures from Benghazi.

What would it be if we could just get one thing done correctly? Obviously, the past is the past, and today, we want action…here it is! When it comes to all these terrorists that attacked us, we request action by the U.S. Government to do everything in its power to arrest the living attackers, especially those easily accessible in Europe. This includes adding bounties to their heads to incentivize reporting and foreign government action on them. Further, a number of the attackers are in jail; while we support the legal process the LNA has against terrorists in its custody, there is the potential for LNA prisons to be attacked and prisoners freed. Access to Benghazi attackers in LNA custody should be ranked as a medium priority.

On the other hand, we lack confidence that any of the terrorists held in the custody of the Government of Libya will remain in custody permanently. Access to Benghazi attackers in Libyan Government custody should be ranked the highest priority. As a starting point, any Benghazi attacker not already freed from Mitiga Prison in Tripoli should be considered for extradition to the U.S. to face justice. It is unclear why there have been no requests to extradite them until this point.

Separately, attackers in hostile zones are prime for targeting actions. A key target for action is Hashem Bousidra. While the LNA had been at the forefront of successful airstrikes against the Benghazi attackers, the U.S killed seven attackers in airstrikes inside Libya (see following graphic). However, in none of the instances was it reported publicly that terrorists involved in the Benghazi attacks were killed.

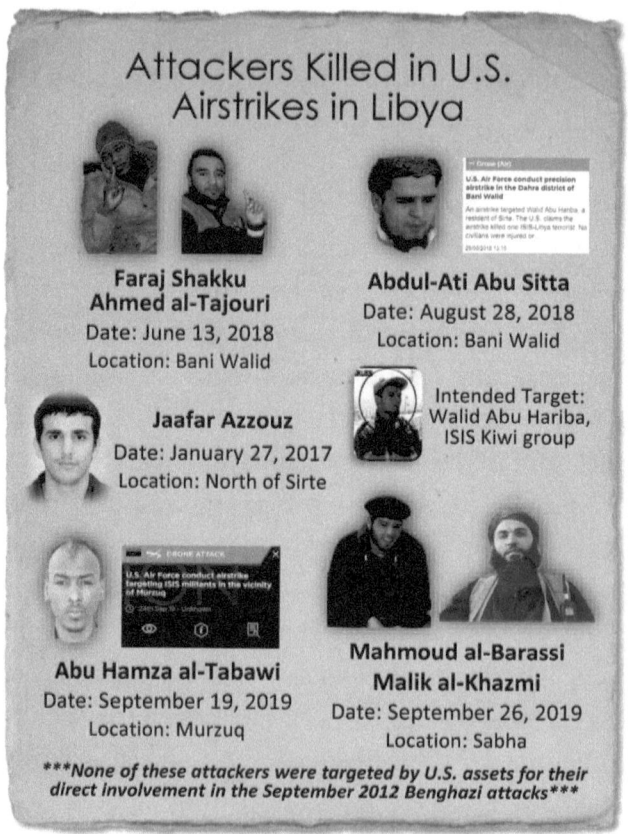

The FBI has also never gone after anyone who attacked the CIA Annex in Benghazi on September 12th, 2012. The two "attackers" they did detain showed up late to the U.S. Consulate attacks and essentially looted the compound. As the attack on our Annex was glaringly ignored, many believed the Libya Shield attackers were our rescuers that night. These attackers have even been written up almost as heroes in official Congressional reports when they blew us up with mortars! And remember the timing when they chose to fire mortars at us—after al-Qa'ida had already caused the death of our Ambassador. Wissam and his deputies in Libya Shield did not care. They went in for the kill, they wanted to expel us from Benghazi that day, and they successfully did so without repercussions.

As noted above, it was the FBI and the CIA, including our infamous COB Bob, who relied on Wissam Bin Humid to provide security for them when they returned to Benghazi in October 2012 for the first time to collect "evidence" related to the attacks. The lead CIA Counterterrorism Case Officer from Benghazi, who garnered the underground former Gaddafi-affiliated Libyan Military Intelligence rescue force the night of the attacks, protested this move. The Officer noted at the time that it was too risky to trust Wissam as his role in the attacks had yet to be thoroughly investigated, and a decade on, it never was! Wissam played the FBI and the CIA for fools, and they let him as Benghazi was never about true justice.

Politicians turned our attacks into political talking points to affect an upcoming Presidential election. This made those in the Intelligence Community, who honored their oath to remain apolitical in their assessments, not want to touch the case or work on efforts against the attackers. Guess what!? If you are a politician who wants to put Americans in war zones where terrible things happen to good people, at least do not let your ambitions ruin their opportunity to seek justice. Those we lost deserved more from our government before, during, and after the attacks—for honorably serving their country and for losing futures with their families and with us, their friends.

The three dozen plus of us from Benghazi have stayed relatively quiet. We have asked very little of our government, but so many more people have unjustly died because not enough was done against the attackers in this book. Almost all of the attackers got away with their involvement in Benghazi. Even those killed, their role in Benghazi was hidden, until today.

Yes, al-Qa'ida paid attention to how easily they got away with the attack on a U.S. Consulate and the killing of one of our generation's most beloved U.S. Ambassadors. We continue to underestimate our enemy entirely. Luckily, jihadi life is tough. Thankfully, men like General Haftar and his army exist. They took on this counterterrorism battle and gave us a masterclass in how you end terrorism. A struggle that a coalition of first-world countries, with unlimited coffers and clearly unlimited time, hadn't successfully handled in places like Afghanistan. Our government chose to deflate the ball so they didn't have to play the game to bring our attackers to justice. And everyone now knows that, not just us!

Acknowledgements

Acknowledgements

We want to give a heartfelt thank you to those who have supported this mission and who have sacrificed to bring true justice to the attackers of our friends in Benghazi; We all chose to serve in Benghazi because we believed in a free Libya—especially free from terrorists

First off, we would like to honor our brothers who we served with and who sacrificed both for the State Department and the CIA on the ground that evening and for the protection of the United States against its enemies abroad. And we thank their families for raising them to be the type of men that it was an honor to serve alongside. There's always the wish to go back to that time and have just a few more moments with them. In memory, those include:

Ambassador Christopher Stevens
Sean Smith
Tyrone "Rone" Woods
Glenn "Bub" Doherty

We also recognize those lost in In Amenas from January 13th to January 16th, 2013, as it was our failure as an Intelligence Community to stop the attackers after Benghazi, which gave them agency to do worse

Acknowledgements

in Algeria. In memory, those Americans include:

Frederick Buttacio

Gordon Lee Rowan

Victor Lynn Lovelady

We would also like to honor all those Libyans who have fallen since 2011 after being impacted by the terrorists in our investigation. We fought these attackers one night, but we understand the sacrifices you and your families made to end their reign, especially in Benghazi. We honor those who fought first against the tyranny of Gaddafi and then against the tyranny of these terrorists.

We thank our CIA colleagues in Tripoli, who were forward-leaning to ensure assistance was sent to Benghazi. We also would like to thank our State Department partners, posted in Benghazi, an extended family to us at the time. We appreciate your efforts to help defend our CIA Annex during the three attacks on us on September 12th, 2012.

To the good Samaritans who found Ambassador Stevens in his villa, carried his body from the building, and ensured that he safely made it to the Benghazi Medical Center, we appreciate your efforts to save him. He loved the city and the people of Benghazi, and while it will always haunt us that we couldn't find him through the smoke, we find peace in the fact that it was good Samaritans that were the last persons in Libya to spend moments with him.

Of most importance, we would like to thank the Libyan National Army (LNA), its General Khalifa Haftar, and all the personnel who have served the LNA over the years for bringing real justice to so many of the Benghazi attackers. We thank you for all the terrorists you've removed from the battlefield—the ones in this book and within the terrorist networks that enabled them. We also thank your partners to include the Egyptians and the Emiratis; in particular, the United Arab Emirates Air Force (UAEAF), for bringing justice to Wissam Bin Humaid. Thanks to you all, the Mastermind of the attack on our CIA Annex will never commit an attack again.

Personally, for you, General Haftar, honestly, sir, you deserve a whole book focused on your commitment to defeating terrorism and for never easing up on your efforts to target the 2012 Benghazi attackers. No one has helped our fallen rest more than you, nor given those of us who served hope. We commend you for your never-ending loyalty to your country and dogged fight against terrorists—our one true enemy in the region and the one truest enemy of the Libyan people and their future. While the U.S. Government waivers on being true partners to those who share our values, we on the ground will always value your service to your people. As Americans, one of our greatest ideals is patriotism, and we respect the patriotism you hold for your country. It's honorable.

We thank our French Partners for bringing justice to the top three AQIM terrorists involved in the Benghazi attacks. First was U.S. Consulate attack Mastermind MBM in November 2016; then Yahya Abu al-Hammam in February 2019; and lastly, Abdel Malek Droukdal in June 2020. Thank you for ending their reign of terror in North Africa.

To the members of the rescue force who referred to themselves as the Libyan Military Intelligence in 2012. We are not sure what has come of your organization, but on the morning after the attacks, no "ally" of our Government showed the humanity that your underground outfit showed our brothers and sisters. We humbly thank you for choosing to evacuate Americans from the CIA Annex to the Benghazi International Airport—an act that provided no benefit to your organization, nor considered the safety of members of your own service. We will always remember who had our back after the mortars fell. You were warriors, and we know we will never get the opportunity to thank every member of your force who showed up for us that night, but we can promise, you are all never forgotten. You helped give us heroes that night, as well.

And lastly, only last, because there are so many of you—we thank the citizens of Benghazi who are against Islamists. We will never forget the mass protests of 20,000 Benghazi Strong on September 21st, 2012, as you forced terrorists AAS-B from their militia barracks. It's you who are still there; it is you who carried on the fight when we should have been there as your allies. Please keep shouting from the rooftops, "The blood we shed for freedom shall not go in vain!" As that's true, the revolution only belongs to the people! And so does the future of Libya!

ATTACKS TIMELINE

Benghazi Attacks Overview

In Benghazi, six separate terrorist attacks occurred throughout September 11th and 12th, 2012.[v] On September 11th, two attacks were on the U.S. Consulate in Benghazi. The first attack occurred at 2142 local in Benghazi (1542 in the afternoon in Washington DC). It ended at 2214, with well over 100 attackers inside the Consulate and at least 50 terrorists manning the roads adjacent to the Consulate. The second attack on the Consulate started almost an hour later at 2310. A third attack on September 11th was an ambush at 2316 against an armored vehicle full of State Department officers after departing the Consulate.

On September 12th, 2012, there were three attacks on the CIA Annex. The first attack started at 0034 and lasted approximately ten minutes. This initial attack was likely an attempt to probe the security posture at the Annex. Then, at 0110, a much larger force of approximately 20 attackers with heavier, sustained arms carried out a continuous assault for the next half hour. For hours after this attack, there was a lull in fighting and some likely surveillance incidents. The final attack on the Annex and the final of the six attacks on us Americans commenced that morning at 0517.

Just before dawn, six mortars impacted the CIA Annex. While al-Qa'ida's Leader at the time, Dr. Ayman al-Zawahiri, directed the planned attack on the Consulate, the mortar attack on the Annex was a target of opportunity taken by the Leader of Libya Shield One, Wissam bin Humaid.

2100	**SEPTEMBER 11TH—ATTACKS ON THE CONSULATE—PART I**
	State Department Officers Ambassador Stevens and Sean Smith had both retired to their rooms for the night; The Ambassador added to his diary the quote, "Never ending security threats..." The Ambassador retiring to his room set off the subsequent chain of events; Al-Qa'ida laid in wait in the vicinity of the Consulate until they confirmed the exact location of the Ambassador; Al-Qa'ida then commenced an operation to kidnap him
2103	Omar al-Shaalali arrived to serve as al-Qa'ida's Ground Commander for the U.S. Consulate attacks; At the same time, a Supreme Security Committee (SSC) vehicle arrived; SSC then provided cover for al-Qa'ida as Omar spent about 40 minutes conducting surveillance of the area surrounding the compound, and preparing his arriving attackers
2131	Al-Qa'ida in Iraq (AQI) Senior Operative Ali Owni al-Harzi, sent the first of three text messages to AQI associates that he was on his way to attack the U.S. Consulate in Benghazi
2140	**FIRST ATTACK ON THE CONSULATE**
	The vehicle belonging to the SSC departed the front of the Consulate and did not warn any Americans regarding the impending attack
2142	At 1542 EST, al-Qa'ida Commander Omar leads the direct assault on the Consulate, breaching the Consulate initially with over 100 terrorists with the first couple dozen visible on the surveillance cameras at the Consulate's main gate; This included Majdi al-Mushaiti, Ahmed al-Majbari, Ayman Shahat al-Dijawi, Zakaria al-Bargathi, and Mohammad al-Werfalli; Additional terrorists had manning positions on the roads surrounding the perimeter of the Consulate including Majdi's cousin Ahmed al-Mushaiti and

brothers Yunus and Ismail Sakarya; State Department's Diplomatic Security (D.S.) Officer manning the Tactical Operations Center (TOC) hit the Imminent Danger Notification System (also known as the duck-and-cover system); The D.S. officer then phoned us at the CIA Annex noting they were under attack and asked for an emergency response

2143   At our CIA Annex, we could hear explosions and shooting in the background, not uncommon in Libya at the time; We were called over the radio by our Team Leader (T.L.) to come to the CIA's main office, but he did not tell us why, so I assumed we were going to be given a briefing on the events that occurred in Cairo as I had learned about it initially from an email sent to me by Sarah when it kicked off; As we were moving our way to the SCIF, our T.L. called again with more urgency telling us to come, immediately, making us realize that we likely had an issue in our own backyard; Throughout the attacks, our T.L. miscommunicated details of the events on the ground like in this case—he could have just said there was an attack on the radio as it was a public event

2144   Back over at the Consulate, the attackers had breached the compound and set fire to the 17 February villa, which the local armed guards used; Again, members of 17 February also participated in the attacks, including Salah al-Mushaiti, Talal Bin Hariz, Imad al-Awami, and Khaled al-Ammari—Salah, Talal, and Khaled also brought their brothers to participate in the attacks including Hamza and Shoaib al-Mushaiti, Faisal and Hudhayfa bin Hariz, and Muftah and Mohammad al-Ammari; Additionally, members of the Blue Mountain Guard forces were also involved to include Faraj Shakku and a brother of Ahmed al-Mushaiti named Mohammad; This fire gets misconstrued with the fire that is later set to the Ambassador's villa

2145   Over in Tripoli at the U.S. Embassy, the Regional Security Office (RSO) received notification of the attack again from the TOC in

	Benghazi; Our Chief of Base (COB), Bob, calls Fawzi bu Khatif, the Leader of 17 February, asking for the group to respond to the attack on the Consulate; Fawzi denies Bob's request noting he will NOT be sending reinforcements
2146	Over at the CIA Annex, on CCTV our GRS team is seen putting on armor then loading up in agency vehicles, where the team remained stationary for almost another 20 minutes
2149	In Washington DC, the Diplomatic Security Command Center (DSCC) received word that the Consulate in Benghazi was under attack
2150	The Deputy Chief of Mission (DCM) in Tripoli connected via cellphone with Ambassador Stevens, who reported the Consulate was under attack before the phone disconnected; Back north of Benghazi's city center, Ahmed Abu Khatallah, who was incorrectly identified as the attack mastermind, was at home having tea with an associate when according to the FBI investigation he received calls from attackers Zakaria al-Bargathi and Aymen al-Dijawi telling him about the attacks; He then arrived at the Consulate eight minutes later
2200	The terrorists sole focus shifts to breaching the Ambassador's villa as their primary mission was to kidnap the Ambassador; The attackers are unsuccessful in accessing the building
2201	Nineteen minutes into the attacks, as the attackers could not access the Ambassador, they set fire to the villa to smoke the Ambassador out; This led the Ambassador and two additional State personnel to abandon the safe room where they had sought refuge; At the CIA Annex, our GRS team is still loaded in our vehicles, but Bob, and our T.L. refuse to let us leave; Even as Bob was fully aware there would be no response to the Consulate from 17 February
2203	The only other individual detained by the U.S. Government for

his involvement in the attacks, Mustafa al-Imam was called by Khatallah when still at home and was notified about the attack according to the FBI investigation; They then planned to meet up, and Imam arrived five minutes after Khatallah where they met up outside the compound on a nearby Venezia Street with the closest entrance being that back gate of the Consulate; They did not enter the Consulate until after al-Qa'ida's attackers departed

2204   A D.S. Officer in the Ambassador's villa reported that he had become separated from the Ambassador and Sean Smith due to heavy smoke; The D.S. Officer in the TOC continued to radio us at the CIA Annex over and over asking for assistance as Bob refused to let us depart—frankly Bob had no intentions of letting us go, he had a long history[vi] of refusing to allow us to respond to incidents in Benghazi as he wanted to rely solely on 17 February, an Islamist group, instead; Finally, the D.S. Officer screamed that they were all going to die, and we knew we just had to leave

2205   Twenty-three minutes after being notified of the attack on the Consulate, we (the GRS team) loaded up to leave, blowing Bob off, and yelling at our T.L. to get in the car; We then departed for the Consulate in two armored vehicles; Back in Washington, DC the White House Situation Room was notified of the attack

2206   After our GRS team departed, the CIA Communications Officer at the CIA Annex realized that neither Bob nor his Deputy had called the CIA Operations Center in Washington, DC, nor had they called the CIA Station in Tripoli to notify them of the attack; So, he took the lead and reported the attack to CIA Headquarters

2207   The U.S. Consulate TOC reported to the Regional Security Officer (RSO) at the U.S. Embassy Tripoli that contact with the Ambassador had been lost; Our CIA team had now arrived in the vicinity of the Consulate; as we dismounted our vehicles at one of the roadblocks set up outside the Consulate

2208	At the roadblock, we were told that it was too dangerous for us to go down the road and that we would need permission from Ansar al-Sharia to proceed, and almost simultaneously, automatic heavy machine gun fire came down the road at us from the front of the Consulate; In response, I told a fellow GRS Officer to fire a few 40-millimeter grenade rounds from a grenade launcher towards the source of the machine gun fire; This show of force proved to be effective, and although we didn't know it at the time this one operational choice was responsible for the initial retreat of the attackers on the Consulate
2210	Bob finally called the CIA's Chief of Station (COS) in Tripoli and notified him of the attack; This delay in communication to CIA Leadership is a critical element to note in the timeline; This delay is why both the COS and senior leadership at the CIA HQS to include then CIA Director Petraeus believed GRS was allowed to and responded immediately to the crisis as when attack details were finally reported up—it was also reported simultaneously that GRS had already left in response; In real-time it was unclear to CIA Leadership outside of Benghazi that the attack had been ongoing for almost thirty minutes before they were even notified
2213	In just five minutes, all the attackers fled off the U.S. Consulate grounds, with al-Qa'ida Commander Omar being the last out among his al-Qa'ida attackers
2214	All the attackers have officially departed; If Bob had allowed us to depart the CIA Annex and not caused a deadly delay, the grenade launcher could have been fired by 2150 (when we should have arrived near the Consulate), which would have potentially prevented the attackers from setting fire to the Ambassador's villa
2215	A detailee based at Tripoli Station requested permission from the COS to reposition the unmanned aerial vehicle (UAV) to Benghazi that was located east of Benghazi monitoring al-Qa'ida

	Senior Operative and AAS-D Leader Sufyan bin Qumo; Qumo sent approximately 15 to 20 terrorists to the Consulate attacks including Ali bin Taher, Hani al-Hawari, Anis al-Khurram, and Abdul Hamid al-Shaeri
2217	The first D.S. Officer comes out to clear around the U.S. Consulate buildings as the attackers are no longer inside the walls of the compound
2220	COS in Tripoli established a CIA-led response termed "Team Tripoli" to respond; His mission for the QRF was to deploy to Benghazi to secure and locate the Ambassador; The COS decided to deploy all but one of his Tripoli-based GRS Officers to Benghazi; While a calculated risk, a bit of reprieve came when a large contingent of Zintani militia forces arrived to reinforce the outer perimeter security, preventing any local threat actors from approaching the CIA Annex in Tripoli
2221	Back in Washington DC, at 1621 EST, the White House Situation Room convened a meeting
2230	Our CIA GRS teams begin to arrive inside the U.S. Consulate, and we worked with our State D.S. partners to clear the rest of the compound and begin looking for the missing State Department officers, including the Ambassador
2232	At 1632 EST, fifty minutes into the attacks, the National Military Command Center (NMCC) at the Pentagon notified the Office of the Secretary of Defense and the Office of the Joint Chiefs of Staff of the attack
2238	Two officials from the Libyan Intelligence Service (LIS) arrived at the compound to assist; They were mischaracterized in the Tripoli TOC log as a militia as we all rebuked questioning by these officials believing they were 17 February as Bob had still been telling us dishonestly that 17 February was responding; For years there has been

	a misinformation campaign that 17 February provided support
2256	The TOC, the key source of information sharing that evening, closes down, and the responsibility for information sharing moved to the CIA Annex; As such, the follow-on attacks were downplayed for the remainder of the evening, and most events do not get reported out of Benghazi at all
2300	Back in Washington DC, at 1700 EST, President Barack Obama is finally informed of the attack by his National Security Adviser at the start of a pre-scheduled meeting at the White House with Defense Secretary Leon E. Panetta and General Martin E. Dempsey, chairman of the Joint Chiefs of Staff; Over 90 minutes into the attacks, during the meeting, President Obama ordered the Pentagon to begin mobilizing all available military assets to respond to a range of contingencies in Libya and other countries in the region; There were no direct orders to respond with military action in Benghazi
2301	**SECOND ATTACK ON THE CONSULATE**
	Part of our team helped find Sean inside the Ambassador's villa, and he is reported as killed in action; I then left the U.S. Consulate compound to retrieve one of our vehicles from the original roadblock; I later transported Sean back with me to the CIA Annex
2305	Al-Qa'ida Commander Omar officially ended the al-Qa'ida attacks for the evening on behalf of AQIM's Leader Abdel Malik Droukdal; In Omar's personal opinion, he viewed the attack a failure as al-Qa'ida was unable to kidnap the Ambassador which meant he was unable to save his brother, former AQI terrorist Adel, who was in Iraqi custody awaiting execution; There is no evidence that Omar or any of his known al-Qa'ida associates participated in the later attack on our CIA Annex
2310	Efforts to continue searching for the Ambassador were halted when

a second attack started on the Consulate; In real-time, we thought that perhaps a "friendly" militia had arrived as we still believed 17 February was responding; Our GRS team engaged in a gun battle with the attackers while the D.S. Agents loaded up in a vehicle and within ten minutes of the second attack, at 2316, five D.S. Agents fled from the main gate of the Consulate under heavy fire and headed west; Our GRS team stayed to repel the second attack as ongoing small arms fire and explosions continue in the vicinity of the back gate of the Consulate

2311 The UAV that had been repositioned from Darnah arrived over the compound; Communications between our ground team and the unarmed drones completely failed; We thought the drone operators were receiving our requests, and we found out after the attacks that they had not received even one of the inputs provided; This appears to have been a failure by CIA Officers in how they communicated amongst each other in chat between Benghazi and Tripoli

2317 **THIRD ATTACK ON CONSULATE PERSONNEL, VEHICLE AMBUSH**

Immediately after departing the Consulate, the D.S. agents reported to the Tripoli RSO that they had encountered a direct ambush by approximately 30 attackers just 200 meters from the gate; The attackers came out of a nearby compound and even attempted to guide the vehicle into the compound

2319 During a break in small arms fire, our GRS team departed the main gate and headed east with Sean's remains as we could not find the Ambassador; We had believed he might have been kidnapped; As it was the opposite direction, we did not meet any resistance; Even after we all left, confusion caused attacks on the Consulate to continue with RPGs, small arms fire, and unknown explosions either with attackers just shooting to shoot or to battle in some way accidentally with one another

2321	Soon after all of our American security personnel departed the compound, a Reuters photographer arrived at the Consulate taking the first pictures to be shared with the world of the attacks' aftermath; The first picture he snapped that night is seen on the cover of this book; The timing of his arrival is important as the attack was now winding down to where local Libyan onlookers and looters started entering the compound
2322	Over the next twenty minutes, the looting of the Consulate begins with the armed and unarmed persons entering through the back gate of the compound; An armored vehicle is looted in addition to one stolen by an attacker during the first wave of the attack, as well as paperwork being stolen from the TOC, and various items taken from inside the Consulate vehicles; As noted, one al-Qa'ida attacker Faraj al-Chalabi reportedly hand carried some of these items from Libya to Pakistan where he intended to give the items to then al-Qa'ida's Leader Dr. Ayman al-Zawahiri, but instead, he was detained by Pakistani authorities
2323	At our CIA Annex, the D.S. Agents arrived thankfully all in one piece with a severely damaged vehicle from the ambush; The vehicle being armored saved their lives as it looked like it had been through war and back
2330	Back near the U.S. Consulate, Libyans reported seeing Libya Shield's Wissam bin Humaid and Khatallah speaking while both were loitering on Venezia Street outside the Consulate; Khatallah was questioning everyone to include even looking at photographs on people's cameras and later detained one of the good Samaritans that pulled the Ambassador out of his villa; Khatallah would finally be seen inside the Consulate at 2354 as a looter
2337	State Department's Political Affairs Officer reported that a Facebook Page called "Tripoli Council" was calling for an unfounded attack on the U.S. Embassy in Tripoli; This is when the little focus there

was in Washington DC and, in particular, the Department of Defense (DoD), on Benghazi was lost with no understanding after this point that attacks were still ongoing in the city

2338 Back over at the Annex, our GRS teams arrived as we took a longer route back to ensure no al-Qa'ida attackers or surveillance followed us; In retrospect, as it was Libya Shield's Wissam Bin Humaid that was the one who attacked us, he was well-aware where our Annex was; All the local Libyan militias did, and many of them even referred to it as "Bob's House;" Back in Washington DC, at 1538 EST, Secretary Clinton made the first call to then CIA Director David Petraeus almost two hours into the attacks

2345 Our own GRS T.L. called off the medical evacuation (medevac) with attacks still ongoing (and with three more attacks to occur, one deadly); This was the only DoD resource requested by the CIA during the attacks, as the COS in Tripoli had requested it; None of us were informed that he had called it off, and we only learned well after the fact; Even as the attacks started on the CIA Annex, we found out our local CIA Officers had downplayed the attacks as they were occurring; For example, our T.L. reported the attacks as only being "pop shots" as he stayed safely in the SCIF building never firing a shot that evening at either the U.S. Consulate or the CIA Annex

2359 The end of September 11th, 2012

SEPTEMBER 12TH—ATTACKS ON THE CIA ANNEX—PART II

As a review, on September 11th, there were three distinct attacks against us, two at the U.S. Consulate and one ambush near the Consulate on a Consulate vehicle; Little had been done in Washington DC at this point in response to the attacks in Benghazi, and the focus by this time had moved off of Benghazi where ongoing attacks continue to Tripoli where there was only the threat of an attack from a sole Facebook post not from a known extremist page; Our Ambassador was still missing, and those of us at the CIA Annex were unaware that we would have to fight through three more waves of attacks before dusk on September 12th, 2012; We don't know in real-time but learn after the fact that no U.S. military assets were directed to Benghazi throughout this crisis, however, we will continue to watch and wait for them

0000 Back in Washington DC at 1800 EST, Secretary Panetta and Pentagon officials engaged in meetings for two full hours until 2000 EST, where options were discussed primarily regarding potentially sending support to Tripoli; Concerns were also raised over the outbreak of violence in the region; Panetta gave a vocal order for military action "to expedite the movement of forces upon receipt of formal authorization;" For the first time, seven hours after the Egypt unrest where 20 of the 2000 protestors breached the U.S. Embassy in Cairo; and over 4 hours now after over 100 attackers overran the U.S. Consulate in Benghazi, the Secretary of Defense is saying "get ready" which did not include formal authorization

0006 At the same time the DoD was meeting, the State Department's Operations Center issued a second "Ops Alert" sent via email to the White House, Pentagon, FBI, and other government agencies noting that Ansar al-Sharia claimed responsibility for the attack; The existence of the email was not disclosed to the public until Reuters reported it on October 24th, 2012

0025 Over in Tripoli, a CIA Staff Operations Officer who was supporting Team Tripoli as a linguist procured a private charter plane from a local Libyan to use as transport to Benghazi that evening from Tripoli's Mitiga airport; In a twist of fate, the CIA Officer had met the individual for the first time earlier in the day, as the CIA was looking into options for air transport in Libya; This Libyan had reportedly been a Gaddafi loyalist during the regime

0030   Back over in Benghazi, Bob received a call from al-Qa'ida Commander Mohammad al-Gharabi, the Leader of the Rafallah al-Sahati Brigade; Gharabi offered for the CIA personnel to come over and hunker down for "safety" at his headquarters; At this location, were seven Iranians detained in July 2012, who while reported to be working for the Red Crescent, but most were intelligence officers from Iran's Islamic Revolutionary Guard Corps Quds Force; Separately, it was Gharabi, along with Wissam Bin Humaid, who on September 9th, 2012, told State Officers in Benghazi that they could no longer ensure the security for Americans; Gharabi also provided attackers for the U.S. Consulate attacks to include Jalal Makhzoum and his long-time al-Qa'ida associate Marei Zoghbi also participated

0032   FIRST ATTACK ON THE CIA ANNEX

As attackers were filing into the areas around our Annex, we GRS members on the roofs of the three-separate buildings asked our T.L. and Bob to call 17 February to determine whether any of these individuals were "friendly" forces and coming to assist; Again, 3 hours after 17 February's Leader told Bob he was not sending assistance, Bob had not still informed us; The CIA subsequently rewarded Bob for his "leadership" that night even as he intentionally put all our lives at added risk

0034   The attackers carried out their first successful attack against our Annex after staging gun trucks at a parking area adjacent to the northeast intersection; They used the east field as cover and concealment to move toward our exterior wall; Small arms fire struck the northeast portion of the Annex and destroyed the flood light in the northeast corner, according to our surveillance footage

0035   IED and RPG attacks commenced; By the end of the night, attackers had attacked us from three of the four sides of the Annex, to also include a deadly aerial attack

Benghazi Attacks Overview

0041   Sustained small arms continued as explosives impacted the east side of the Annex; Attackers in the east field continued to regroup back to staged gun trucks after taking direct fire from our GRS team and supplemented by D.S. elements on the Annex's tower and roofs; After the end of the first attack, CIA staff personnel were being escorted below us by Agency Physical Security Officers presumably to pack and prepare their go bags; No one was briefed on the CIA plan that night for evacuation by either Bob or our T.L. and most of the GRS on the roofs were not even aware staff were preparing bags to depart the Annex due to the severe lack of communication

0045   There was now a break in the attacks, and a peacefulness fell over the Annex, which we knew would be naive to get accustomed to; Sarah was about to go to bed in a European Capital after leaving Benghazi in the early morning hours of September 11th, and called my cell phone while I was on the roof; I noted I was safe, but busy and said goodnight, not disclosing that a series of attacks had occurred; Jokingly, Sarah had assumed I was busy playing Call of Duty, a favorite past-time, which we referred to in GRS as "professional development"

0059   Attackers remained visible in the east field; However, we refused to fire on their location as it was too close to a residential house where a local family lived, including children; One issue inside the Annex was that our Communications Officer had requested over and over again for Bob to allow him to start destroying classified documents; Bob, a hoarder, refused for three hours; As a result, there were not enough shredders available to timely destroy the mass holdings he had printed against CIA's classified holding policies; Base finally started shredding at 0100, and the CIA redacted the time it took to shred Bob's collection of classified documents for this publication

During this first attack on the CIA Annex, Secretary Clinton made her first call to Libyan President Mohammed Magarief

0100   Back over at the Consulate, a curious local found the Ambassador on the ground in a bedroom; He ran outside and requested assistance from a young Libyan Army Officer; The two good Samaritans pulled the Ambassador out and laid him down briefly on the front porch; They celebrated with other onlookers and proclaimed "God is Great" in Arabic believing the man they found was alive; This was an incorrect assumption as three hours had surpassed since the fire was started

0105   The locals then carried him towards the Consulate entrance to seek transport and were ushered to a car belonging to a neighbor who transported the Ambassador to Benghazi Medical Center; Remember, these were good Samaritans; They meant no ill will; His body was never mishandled; They put forward their best efforts in an attempt to save him, and we thank them for their humanity

0110   After arriving at the hospital, because of the soot covering his face, the Doctor initially failed to recognize the Ambassador as everyone in the city of Benghazi knew him due to his efforts to protect the people of Benghazi during NATO's intervention

0111   **SECOND ATTACK ON THE CIA ANNEX**

After the first Annex attack, we knew the attackers would regroup and attempt another assault; The attackers did not expect heavy resistance, as al-Qa'ida over at the Consulate had not and had prepared for it by co-opting local guard force members; Our Annex was much better prepared for a firefight; the Annex was also more allowing for 360-degree combined views

Realizing our team would come down off that adrenaline high, causing their bodies and minds to crash, I went and grabbed some candy and Gatorade to pass around and keep energized; This was the first and only time I left the rooftop until daylight when I was the last to drop my field of fire, and overwatch position when the Annex eventually evacuated

0112   I was back on the roof when the second attack kicked off with an RPG strike and then sustained small arms were fired from the east side of the perimeter wall; Two cameras were destroyed, and another possible RPG struck the Annex from the east field, limiting surveillance footage for the rest of the evening; The last of the roof battle commenced from the roof I was on top of; After the firing stopped, an eerie quiet came over our complex

0115   Khatallah is still lurking about the Consulate when a second attack is occurring at the CIA; And as Khatallah loved inserting himself in an attack he was barely involved in, he captured the good Samaritan Libyan Army officer and questioned him regarding the attacks

0119   Back in Washington DC, Secretary Panetta's Chief of Staff, Jeremy Bash, emailed leaders at the State Department to inform them of the assets that could be deployed in response to the attack; "After Consulting with General Dempsey, General Ham, and the Joint Staff, we have identified the forces that could move to Benghazi; They are spinning up as we speak; They include a SOF element that was in Croatia (which can fly to Suda Bay, Crete) and a Marine FAST team out of Roda, Spain"

Note that this was the rare time the DoD mentioned responding to Benghazi in communications, as it then shifted to Tripoli or Libya in general for the remainder of discussions, and nothing listed was ever sent to Benghazi

0130   1930 EST in Washington DC, almost four hours into the attacks, the White House convened a secure video teleconference with the State Department and the DoD on responses to Libya; The two representatives assigned by DoD as its points of contact regarding the attacks, Admiral James A. Winnefeld, the Vice Chairman of the Joint Chiefs of Staff and Dr. James Miller, the Under Secretary of Defense for Policy did not even attend the meeting; Winnefeld was not even at the Pentagon as he had gone home to host a dinner party for foreign dignitaries and reportedly only had one discussion regarding any updates to the attacks during the dinner; Reportedly, after his dinner concluded around 2200 (0400 local in Benghazi, after five direct attacks against us had occurred), he then went to the secure communications facility in his home and "checked emails"; Miller was not at the Pentagon either due to an unexpected family emergency; He asked Bash to fill in for him during the White House meeting

0135 Over at the Benghazi Benina International Airport, Team Tripoli landed during the lull in fighting at the Annex; The DoD members had believed a contact of theirs had set up transport through the Libya Shield, but when the team arrived, no vehicles or reinforcements were waiting for them

0145 Pan-Arab Stations carried a joint news conference by the Libyan President Magarief and Libyan's Prime Minister Abdurrahim el-Keib in which they both condemned the attack on the Consulate; even as their government had financially supported many of the terrorists involved; They also reported that elections for the head of the government would be held on schedule that day, September 12th

0148 Back at the CIA Annex, a vehicle that we believed was surveillance was seen east of our location, and then drove by the Annex, then u-turned and drove past again; During this lull, multiple individuals remained hiding in a sheep heard in the field north of the Annex; Cell phones were being used likely as ongoing surveillance; Team Tripoli was still stranded at the airport trying to secure transport

0150 Reuter's news outlet, citing Libyan government sources, reported, "An American staff member of the U.S. Consulate in the eastern Libyan city of Benghazi has died following fierce clashes at the compound"; This was the first reference to the death of Sean Smith in the open press

0150 Khatallah released his detainee, the good Samaritan Libyan Army officer, and then wrapped up his evening of loitering, looting, and playing bad cop to return home

0200 Secretary Panetta convened meetings with senior officials, including General Dempsey and General Ham; They reportedly discussed "response options for Benghazi" even as none of their final options listed included "Benghazi"; They also discussed the potential outbreak of further violence throughout the region, particularly in Tunis, Tripoli, Cairo, and Sana'a

During these meetings, Secretary Panetta authorized the following: Two Fleet Antiterrorism Security Team (FAST) platoons stationed in Rota, Spain, which were sent to deploy to Tripoli; the EUCOM Commander's In-extremis Force (CIF) (which took 22 hours to get from Croatia to Greece); and a Special Mission Unit (SMU) out of Fort Bragg to deploy to an intermediate staging base in southern Europe; Actions are "verbally" conveyed from the Pentagon, formal authorization was still not given

0203 Now that he was finally free, the good Samaritan Libyan Army Officer, started calling phone numbers in the phone he found with the Ambassador to notify contacts that the Ambassador was brought to the hospital; The calls were assumed to be an entrapment operation as not everyone believed the Ambassador had been left on the compound—some had thought he was kidnapped; The phone ended up being the work phone for the D.S. Agent who had gotten separated from the Ambassador

Back in Washington DC, the FBI notified the Diplomatic Security Command Center (DSCC) that they were opening an investigation into the attack

0205 To run down the hospital lead, the State Department asked Bubakar Habib, who supported the Cultural Affairs Office, to have someone check the hospital; Former 17 February co-founder Mustafa al-Saqzli, the brother of Benghazi attacker Khalid al-Saqzli, was sent as a representative from the Libyan Ministry of the Interior (MOI) to the hospital

0210   Still stuck at the airport, Team Tripoli is notified that the Ambassador was brought to the hospital, likely the Benghazi Medical Center; they received erroneous information that AAS-B was in control of the Benghazi Medical Center when in fact, it was the al Jalaa Hospital; Team Tripoli also received erroneous information that the Ambassador's cellphone had been geolocated within the vicinity of the Benghazi airport; Several other issues regarding locations were misreported that night within CIA circles; Another issue was that the attackers possibly fiddled with a personal tracking device after stealing one of the armored vehicles from the Consulate compound which initially sent out locational information

0217   Khatallah arrived home to the al-Laythi neighborhood, which is over three miles away from both the Consulate and CIA Annex; He remained at home until 1023 in the late morning hours; As such, in addition to showing up to the Consulate attacks late, we found no evidence that Khatallah traveled near the Annex or played any role in the three attacks on the CIA Annex

0230   The DoD conducted a Benghazi Conference Call with representatives from AFRICOM, EUCOM, CENTCOM, TRANSCOM, SOCOM, and the four services; No known actions came of this meeting

0237   Over in Tripoli, the U.S. Embassy Tripoli called an Emergency Action Committee (EAC) to determine if the Embassy was safe; It was determined that all Embassy personnel would relocate to the CIA Annex in Tripoli, which they did at 0258

0239   Secretary Panetta ordered the National Military Command Center (NMCC) to transmit formal authorization for two FAST platoons to deploy and for the EUCOM CIF to separately deploy to a staging position in southern Europe approximately five hours after the start of the attacks

0253	Secretary Panetta ordered the NMCC to transmit formal authorization to deploy a Special Mission Unit (SMU) from the U.S. to a staging position in southern Europe
0322	Back at State Department in Washington DC, at the DSCC, the Senior Watch Officer emailed out the first official notice regarding the death of the Ambassador as follows, "Embassy Tripoli confirms the death of Ambassador John C. (Chris) Stevens in Benghazi;" At the time, Stevens was the eighth Ambassador to perish in the line of duty, and the first to be killed while serving in office since the mysterious plane crash that killed U.S. Ambassador to Pakistan Arnold "Arnie" Raphel and Pakistani President Zia ul-Haq in 1988
0325	Over at the Benghazi Airport, Team Tripoli was notified of the death of the Ambassador; The CIA's COS in Tripoli told them to travel to the Annex and evacuate non-essential personnel; This would include transporting 14 persons back to the airport who would then take the chartered plane back to Tripoli that Team Tripoli had procured; Initially, there was no second plane to assist in evacuations; This plan, as intended by the COS, would then leave primarily only GRS and security related officers to defend the CIA Annex; Bob nor our T.L. in Benghazi told us GRS personnel in Benghazi of this plan
0328	Denis McDonough, the then-Deputy National Security Advisor, talked to YouTube about removing the Internet video, Innocence of Muslims, which played no role in the planned al-Qa'ida attack on the U.S. Consulate in Benghazi
0415	In an odd side event in an already hectic evening, a U.S. Embassy D.S. Agent now located at the CIA Annex in Tripoli learned that two American contractors would need to be evacuated from Benghazi—no one on the ground in Benghazi was aware of the fact that they were even located in Benghazi; So, another two were added to an already large number of Americans that needed to be

	rescued from Benghazi that night, and remember the U.S. military never planned nor were sending any assets to Benghazi
0425	In Washington DC, the first comments by U.S. Government officials regarding Sean's death came from then Presidential candidate Mitt Romney; In response to Romney's statement, President Obama called Secretary Clinton
0432	@StateDept tweeted: " #SecClinton: I condemn in the strongest terms the attack on our mission in Benghazi today. http://state.gov #Libya" and "#SecClinton: We have confirmed one @StateDept officer was killed in #Libya. We are heartbroken by this terrible loss."
	Secretary Clinton further issued a statement, saying, "I condemn in the strongest terms the attack on our mission in Benghazi today; As we work to secure our personnel and facilities, we have confirmed that one of our State Department officers was killed; We are heartbroken by this terrible loss; Our thoughts and prayers are with his family and those who have suffered in this attack;" She added: "Some have sought to justify this vicious behavior as a response to inflammatory material posted on the Internet; The United States deplores any intentional effort to denigrate the religious beliefs of others. Our commitment to religious tolerance goes back to the very beginning of our nation. But let me be clear: There is never any justification for violent acts of this kind"
0427	Twitter broadcasted a photo of Ambassador Stevens pulled from the video of the rescue by good Samaritans before he was transported to the hospital
0434	At the Benghazi Medical Center, Ambassador Stevens was confirmed deceased after Mustafa provided an identification; He then arranged through Libya's MOI for the transportation of the Ambassador's remains to the Benghazi airport
0439	State Department's Undersecretary for Management Patrick

Kennedy (one of the senior officials who refused the almost 600 security requests sent from the U.S. Consulate to State Headquarters in the lead-up to the attacks) forewarned the aforementioned Twitter photograph of the Ambassador to Secretary Clinton's Chief of Staff, Cheryl Mills; Later, it was Cheryl who eventually notified Secretary Clinton of the Ambassador's death via email at 0538

0445 Team Tripoli finally departed from the Benghazi airport after the Libya Shield Two Captain Fathi al-Obeidi received approvals to take the team to the CIA Annex; The person who had been preventing the team from leaving the airport was the Leader of Libya Shield One, Wissam Bin Humaid as a CIA Officer was told they could not depart until Wissam gave approvals; Wissam's close friend, Boka al-Oraibi was Fathi's boss and the Leader of Libya Shield Two; It is assessed that Wissam and Boka worked in concert against CIA's rescue efforts while Fathi was unwitting

0500 Back at the National Command Authority, the DoD N-Hour was established, which finally started DoD preparations to send assets to either Tripoli or Sigonella, Italy, again with no plans for anything to travel directly to Benghazi even after five attacks, not knowing how many additional attacks were incoming, nor if our situation would devolve into a hostage crisis

For example, the 2008 Mumbai, India attack lasted four days with only ten attackers from Lashkar-e-Toiba (LeT); the 2009 attack on the Pakistani Military headquarters in Rawalpindi, Pakistan lasted 18 hours with only ten attackers from Tehrik-e-Taliban Pakistan (TTP); and even after Benghazi in 2013, the In Amenas oil facility attack, in Algeria (which we noted had shared 12 attackers with the Benghazi attacks) lasted three days with 32 attackers led by AQIM's MBM; In comparison, the initial attack on our Consulate consisted of over 150 attackers, more attackers than all three of the incidents mentioned above combined at 52 attackers

0505	**THIRD ATTACK ON THE CIA ANNEX**

The Libya Shield Two motorcade with a dozen vehicles and Team Tripoli arrived at the Annex; The militia members remained with their vehicles while their Commander entered inside with Team Tripoli; When the GRS Team Leader (T.L.) for Tripoli arrived, he was surprised that the Annex was not prepared to evacuate as the directive had never been passed down to us by Bob nor our Benghazi T.L.; Bob had no intentions to leave, and our T.L. gave us the impression that Team Tripoli was arriving to reinforce us, and that did not happen; Only one member of Team Tripoli, Bub, joined us in a fighting position on the roof while the rest went and hung out in the CIA's main office

Our T.L. had also not properly briefed Team Tripoli about the threat, calling the first two attacks on the Annex "harassing fire" and "sporadic" shooting; Due to all the dysfunction, the GRS T.L. from Tripoli took over control of the Consulate as it was evident that Bob could not decide as to his welfare, let alone his people; The Tripoli T.L. then started to evacuate some non-essential State personnel from the Annex roofs in quick succession

0516 Wissam Bin Humaid called the Libyan Shield Two militia members posted outside the CIA Annex and ordered them to return to their headquarters immediately; One of the personnel noted to Wissam that Captain Fathi was still inside; Wissam commanded him to abandon their positions and immediately return to their base on the other side of town without their Commander; On Annex surveillance footage the militia was seen loading into their vehicles and quickly departing

0517 Only ten minutes after the arrival of Team Tripoli, an attack started from the western side of our Annex, which included PKM machine gun fire and RPG attacks; This is the first time the entire evening that the attackers had struck us from this direction, and then the mortars came from above with six mortars impacting our Annex for only 73 seconds; To be exact:

0517.40	1st mortar impacted the north perimeter wall
0518.01	2nd mortar impacted against or just inside of the north perimeter
0518.21	3rd mortar impacted the interior of the Annex on the top of the CIA Main office, which housed the Sensitive Compartmented Information Facility (SCIF)
0518.32	4th mortar impacted CIA SCIF
0518.40	5th mortar impacted against the exterior of the north perimeter wall
0518.53	6th mortar impacted CIA SCIF
0520	After the mortar strike that took Rone and Bub from us, we had no idea if more mortars were incoming, nor whether the attacks had finally ended; It may have been a blessing in disguise that Captain Fathi was stranded with us as attackers all night made attempts to not kill Libyans; At this time, we could not posture for follow on attacks as a bigger issue was keeping the wounded alive; especially as one D.S. Agent had significant wounds; Rone and Bub were also both CIA medics stationed in the country, one being ours in Benghazi and one being Tripoli's; Thankfully, many of us in GRS have advanced life-saving training, and a fellow GRS officer from Tripoli heroically saved the limbs of the D.S. agent and the arm of a fellow GRS officer
0530	After the mortars hit, Bob asked the Counterterrorism Case Officer in a state of delusion to stay behind (while in a building that took three direct mortar strikes with two dead on the roof); The T.L. from Tripoli had to forcibly removed Bob from the Annex when the time came for an evacuation; Bob also refused to allow any classified equipment to be destroyed until after the mortar attack as

he did not intend to leave the Annex; Hence, the Communications Officer had minimal time to destroy equipment before evacuation; During the FBI's one day, site visit in October, all of the CIA's sensitive equipment had been preserved in the SCIF; It was lucky for the CIA that the Annex had not been overrun as plenty of classified documents, communication gear, and computers were left behind

0606 A Libyan Military Intelligence (LMI) 30-vehicle motorcade arrived to evacuate the Annex; This rescue force was found by complete luck; The Counterterrorism Case Officer in Benghazi contacted a police officer who worked an intersection just north of our Annex to ask for assistance in an evacuation; This contact noted that the police did not even have weapons let alone the resources to conduct an evacuation, but that he had a contact in the LMI who may be of assistance

After the police official put these two in contact, the LMI Commander agreed to come and evacuate our personnel; In a twist of fate, the LMI was an underground intelligence organization of Gaddafi loyalists; State Department still incorrectly briefs that the Libyan Shield (the real Annex attackers) saved Americans that night; Ironically, Gaddafi's people were the only group in the city that came to the aid of Americans; LMI was never formally thanked by the U.S. Government; When General Petraeus came to visit Tripoli in November 2012, the same Counterterrorism Officer had asked him to put a request in for the U.S. Government to provide a formal show of thanks, but it never happened; It is unclear what happened to this LMI unit

0634 CIA staff members and D.S. agents evacuated in CIA vehicles with the LMI motorcade while our Communications Officer was inside our SCIF attempting to destroy the SCIF

0637 Then all that's left is me, another GRS officer, Bob, and the CIA

Communications Officer as the last four Americans to depart the CIA Annex in a pick-up truck; We were left by everyone else as the Communications Officer was setting off the device that was supposed to destroy the office; I looked back at our Annex as we drove away not knowing that we'd never go back and get the terrorists who did this to us; Many persons over the years told us they got the same answers when asked when we'd go after the Benghazi attackers... "it was not the right political environment"

We sped up and caught up to the rest of the convoy; One extra sad thing, though, is that we did not realize that Bob had not properly informed the CIA Staff regarding the deaths of Bub and Rone, so the Communications Officer got into the truck all excited to get the hell out of dodge, as he was unaware of their deaths until he saw four bodies at the airport

On the morning of September 12th, 2012, we loaded our injured onto the private plane procured by the CIA—It was the CIA's Chief of Station in Tripoli who was the only American to step up and send any show of force or assistance to Benghazi. We did our part too, and left no one behind.

"When we quit thinking primarily about ourselves and our own self-preservation, we undergo a truly heroic transformation of consciousness." - Joseph Campbell

Endnotes

i   The U.S. Consulate in Benghazi Compound was located in an affluent neighborhood in Benghazi called Western Fwayhat. This area essentially served as the diplomatic enclave in Benghazi. The ad hoc British Consulate had also been located in this neighborhood before closing after the June 2012 attack against the British Ambassador in Benghazi. The Consulate was located just north of Fourth Ring Road (also called Venezia Road), with the nearest main intersections to the east and West being Shari' al-Andalus (also called "Gunfighter" Road by us at the CIA) and Shari' al-Qayrawan Street (also called "Adidas" Road by us at the CIA), respectively. The gravel road just north of the compound was unnamed and was the location of the main gate (also known as the Charlie-1 (C-1) gate), and a second gate (also known as the Bravo-1 (B-1) gate was located to the northeast of C-1. The compound's back gate (also known as the Charlie-3 (C-3) gate) was located on Fourth Ring Road; however, that gate was used for emergencies only and had an armored vehicle blocking it for added security. At the time of the attacks, the compound covered 8 acres after two residential compound grounds had been combined to form one large area with a small alleyway connecting the two original spaces. This was down from 13 acres starting in August 2011 when the compound connected to another building called Villa A, which the State department discontinued use in January 2012. Four main buildings were located on the compound, which included:

(1) the main building, known as Villa C, where the diplomatic office for the mission was located. The Ambassador resided during his trip to Benghazi in September 2012 in Villa C, and the acting Principal Officer would normally reside in Villa C when in town. This is also the building where the safe haven was located that the Ambassador, Sean Smith, and a DS Agent retreated into once attackers breached the compound.

(2) East of Villa C, on the other side of the alleyway, was the Villa Barracks or Villa B (also referred to as the "Cantina" by DS Agents); this building served as living quarters for the DS agents, had a recreation room, and housed the compound's dining facility.

(3) Just south of Villa B, also across the alley from Villa C, was the Tactical

Operations Center (TOC) which was staffed at all times by at least one DS Agent who monitored what was happening in and around the compound and who to had a direct link when necessary to counterparts at the Regional Security Office (RSO) at Embassy Tripoli.

(4) The last building on the compound was in the far northwestern corner, and it was used by the February 17 militia as the headquarters for their quick reaction force on the compound.

ii The CIA Annex was located ½ a mile in a straight-line southeast of the U.S. Consulate Compound. The driving distance between the two compounds was 1.2 kilometers (approximately ¾ of a mile). The main gate was located on the road to the south, which we referred to as "Annex Road." A driveway through the compound connected to a second entrance gate located on the east side of the Annex; however, that gate was used for emergencies only and had an armored vehicle blocking it for added security. The compound covered 2 acres, a much smaller footprint than the Benghazi Consulate compound. Four main buildings were located on the compound, which included:

(1) the northernmost building on the compound, referred to as Building

C, housed the main office. The main office was located in a room just left inside the building's entrance that was converted to a secure area commonly referred to as a Sensitive Compartmented Information Facility (SCIF). The building also had a separate room used as the main office for our GRS team and a third room used by the CIA's Physical Security Officers. The last rooms in the building housed the two senior leaders on the compound, the Chief of Station and his Deputy. Of note, Building B was the building that was directly hit by three of the six mortars shot at the compound during the third and final attack on the compound in the early hours of September 12.

(2) The two buildings to the southeast and southwest of Building B were referred to as Buildings B and D, respectively. Both these buildings were used as housing for the CIA officers located on the compound. When Ambassador Stevens visited the CIA Annex on September 10, a meeting was held with all the staff in the common area of Building D. Then he moved to the SCIF in Building B with a smaller number of CIA officers for a classified counterterrorism briefing. Of further note, the first two attacks at the CIA Annex were directed at Building B.

(3) The last building, closest to the main gate, was Building A, which was also used as housing and was the location of the cafeteria, where meals were prepared three times a day for the officers located on the Annex.

iii Just over two months after arriving, I received a request from, you guessed it, CIA HQS to capture al-Qa'ida operative Abu Anas al-Libi. At this point, we were over a month into attempting to capture MBM. We had tried everything, including kidnapping (well, more like holding hostage) some CIA technical officers out of Europe. We even set up a checkpoint searching every vehicle for a short man with one eye that might humorously smell like Marlboro cigarettes. We never caught MBM, but gave a gallant effort.

While keen on finding MBM, an elusive target, Abu Anas was a different issue. I discussed the request with our Senior CIA Officer as I was temporarily assigned as the Targeter. We were the two primary officers working on counterterrorism as most staff in the station were political officers. We knew exactly where Abu Anas was and assessed he'd be an effortless capture as he was hardly mobile due to advanced illness. We discussed how we both wanted to seek justice for the attacks against our Embassies in Kenya and Tanzania, in which he played a key role. The ethical hurdle was that we were both convinced he would die in our custody, and neither of us believed that was justice.

So, while it was a tough decision, we sent back to our HQS at the CIA that we would be declining a capture operation against Abu Anas due to his advanced health issues. Immediately after sending it, I received an instant message from the Deputy Head of the counterterrorism analytic shop that covered the region. The message stated that I was not a "patriot" for denying the operation.

After we both were long gone, on October 5th, 2013, Abu Anas was captured by the U.S. Army Special Operations Forces, and upon being captured, his health issues were a surprise. Even though it was well reported that Iran had released him on humanitarian grounds, essentially so he could return home to Libya and die accompanied by his family. As expected, he died in custody on January 2nd, 2015, before being tried. A colleague told me they even gave him a liver transplant in New York City, trying to keep him alive.

iv This is an important place to insert a comment as to our investigation. In this book, the terrorists are confirmed as being located at the Consulate sourced to our own open-source investigative efforts. We chose not to use U.S. government reporting as we did not want it to bias our findings. Since we did our research from scratch, items will vary from the U.S. government. Little things like we actually have Sufyan Bin Qumo's last name and do not incorrectly use "Abu" in front of his name; to bigger things like some of the persons named as members of Ahmed Abu Khatallah's battalion in 2011 were never members of his battalion. In many instances in Khatallah's trial, it appeared like he was being tried as if he were Wissam bin Humaid. Even there was a mention of Khatallah going off to collection mortars in the trial, however, when I was in that same meeting with the FBI and the source of the information, mortars were never mentioned. Also, it did not happen anyway, as you will see in the timeline, Khatallah was at home when all the mortar planning was underway for the attack against the CIA Annex.

So back to terrorist Salem Darby. His name was listed in a fabricated CIA intelligence report naming him as being involved in the attacks alongside Khaled al-Saqzli (the U.S. government spells his name as Saglusi); however, we confirmed both of their direct involvement separately. In terms of the fictitious report, not only did it name Darby, it pushed the false protest narrative as the Case Officer just put the information in there even though the source was not from Benghazi nor was he in the city at the time to witness a protest. I know first-hand that the Case Officer fabricated the entire report as I met with the asset. I attended his next meeting since the Case Officer was on vacation, and a TDY contractor who attended the prior meeting had also departed the country. So, it was me, DCOS (aka "Frat Boy"), and an acting Chief of Operations (COPS) who had just arrived in the country that I worked with for a short time in South Asia.

Here's the story: I should have known the meeting would go downhill fast when I met with our GRS Team Lead (T.L.) in Tripoli, who had been the same T.L. that led Team Tripoli. I asked him what high-threat meeting protocols were in place for the meeting with this individual, who was still not an approved asset. He noted that they never handled this asset using high threat meeting protocols, and why would they? When I told him who the asset was, he was shocked and said the Case Officer had lied to him regarding his identity and the threat he posed. I said he was top threat when compared to similarities with the Khowst bomber. Luckily, he was one of the cadres' best GRS government staff members, so I could always trust him to do the right thing. So, once we got the meeting set up with proper CIA protocols to protect all involved, we held the meeting with him.

The meeting started rocky, and we never recouped by the end, as we were naturally asking questions in response to his last report regarding the Benghazi attacks. The asset was so confused and unable to answer basic questions, noting that he did not know anyone directly involved in the Benghazi attacks. The CIA interpreter finally had to stop the meeting and talk to us on the side. He said why are you asking these questions and providing misinformation to the asset. The linguist noted it made him uncomfortable as he felt we were putting incorrect words in the asset's mouth. So, I told the linguist how the report mentioned several attackers as being involved in the Benghazi attacks.

The CIA interpreter, who had also attended the prior meeting, said none of the information in the report was correct. He explained that the Case Officer and the TDY Officer directly asked the asset those specific names. The two then asked the asset if they were involved in the attacks, to which he replied that he had no idea. The two then asked him if he had seen any of the persons they named since the attack occurred. He responded yes and provided physical details. This somehow led to more incorrect information in the intelligence report noting these attackers had changed their physical appearance since the attacks. The report was only four days after the attacks, so there was not much time to grow a beard, for example.

After the meeting, I immediately requested the intelligence report be recalled, which is standard practice at the CIA when inaccurate information is disseminated to the Intelligence Community in the form of a serialized report. Even though they both were present in the meeting, the acting COPS and Frat Boy would not allow it to be recalled saying we weren't there first-hand to hear what he said last month. Remember, our own CIA interpreter was there and said it was also incorrect. I saw them in the meeting; they knew the information was fake, but as that intelligence report got a lot of play, they did not want to make Station look bad. Sometimes, there's a little too much CYA at the CIA. As I failed to recall the report, I wrote up all the specifics of the meetings and put it into a CIA operational cable that accompanied the asset meeting. What I did not know until years later, before the release of the cable, was that someone senior at Station had deleted the entirety of the contents regarding the fabricated intelligence report. Sadly, this won't be the only time involving the attacks that information was removed or not included in CIA cable traffic. Still, it's against CIA policy, and it was a leadership failure to protect the two unethical officers, which sanctioned them to continue this behavior.

v   While we included pertinent items from the timeline to tell our story of the attackers, this is not an all-encompassing timeline of every event that occurred on September 11 and 12, 2012.

vi As a sidebar, to effectively understand the timeline of events that occurred in Benghazi on September 11 and 12, 2012, it is essential to understand that several delays occurred during the attacks. As they all get lumped together, this section will parse them out and explain the factors contributing to each.

Two separate standdown events get convoluted when discussing responses to the Benghazi attacks on September 11th and 12th. There is incorrect reporting regarding items like fighter jets being told to stand down; however, the Secretary of Defense never even discussed nor considered using fighter jets in response, so we are not going to walk through debunking any conspiracy theories.

In the first incident, after the attacks kicked off in Benghazi, four Department of Defense (DoD) personnel at the U.S. Embassy Tripoli, including a medic, had been asked by "Team Tripoli" to respond. As background, the DoD personnel was a team of special operators who had previously served as part of the Site Security Team (SST) at the Embassy. Team Tripoli was the CIA Chief of Station's operational response to Benghazi. However, the team could not gain approval from their leadership to be a part of the CIA's team, so Team Tripoli left without them.

The Under Secretary for Management at the State Department, Patrick F. Kennedy, terminated the SST's responsibilities for the Embassy's security in August 2012. As such, while this team was previously able to travel with the Ambassador, they no longer had approvals to travel with him. During the revolution in 2011, the Ambassador's security detail was ten agents. In August 2012, at the beginning of the month, the U.S. Embassy in Tripoli had 34 security agents, and only 6 were assigned to the Embassy by the end of the month. So, as attacks were increasing throughout Libya, the State Department was significantly decreasing its security posture.

Of all the delays, the most publicized was our delay as a GRS team when departing from the CIA Annex to the Consulate after State Department's TOC called for assistance. To provide context, we did not believe that our CIA Chief of Base (COB), known as "Bob" was ever going to let us leave if we did not finally force the issue. The night of September 11th was not the first time Bob would not let us respond in Benghazi; it just was the first time it led to dire consequences. One must remember that Bob was not allowing us to respond, even AFTER calling the Leader of 17 February, Fawzi bu Khatif, who clearly told Bob that he was NOT sending a response. Bob lied to us all night long as we tried to link up with 17 February like fools putting ourselves at risk.

Let's take a step back; you wonder why I thought we would never be able

Endnotes

to leave the CIA Annex, EXPERIENCE. The following is a listing of six separate situations that occurred in Benghazi where Bob would not allow our GRS Officers to respond to situations occurring where they would typically respond in any other warzone environment:

(1) In approximately March or April 2012, three GRS Officers were chased for a short time in Benghazi by unknown offenders, and they radioed their GRS Quick Reaction Force (QRF) at the CIA Annex in Benghazi for assistance. Bob would not allow the QRF to depart the compound and respond to this request for assistance.

(2) Again, in approximately March or April 2012, the Deputy Chief of Base (DCOB) and a CIA linguist were held up by a local militia at a checkpoint in southwest Benghazi for about two to three hours. QRF attempted to respond, and Bob refused to allow them to leave the Annex. To show contrast, in March 2012, two CIA Case Officers were held up on the main roadway at a militia checkpoint for 30-40 minutes in a regional city nearby. The GRS Officers, including Bub, were able to respond to the incident with no issues raised by the leadership at that location—I know this as I traveled with them.

(3) On June 6th, 2012, an IED detonated and blew a large hole in the Benghazi Consulate's exterior wall. A DS Agent at the Consulate immediately called GRS QRF for backup and assistance. When GRS attempted to respond, Bob would not allow GRS to depart the Annex.

(4) On June 10th, 2012, our closest western ally in Benghazi, the British Ambassador, was attacked, and two of his staff were injured when a rocket-propelled grenade hit their convoy. In response to a request from our British counterparts, GRS tried to respond immediately to the scene, and Bob would not allow GRS to depart the CIA Annex. This caused an extensive delay in response until GRS was ultimately able to leave well after being ready and prepared to respond. This would foreshadow what would occur again almost three months later, on September 11th, 2012.

(5) In approximately June or July 201, there was a standoff with GRS, including GRS Officer Tyrone "Rone" Woods and Ansar al-Sharia-Benghazi (yes, the same group that three months later attacked our Consulate) at a checkpoint

314

in Benghazi. GRS QRF was requested to respond to the checkpoint, and Bob denied their departure from the Annex. In Benghazi, we all looked at Rone as our defacto leader, and being unable to leave to support him was viewed as a betrayal by the staff at the CIA Annex.

(6) In August 2012, Benghazi CIA Staff Officers, including Bob, two GRS Officers, and a State Regional Security Officer (RSO) were held at the Benghazi Airport for approximately five hours. GRS QRF attempted to respond, and to quote Bob, "17 February acts as the QRF for the Annex, and he would never allow them (GRS QRF) to respond to an incident in town." Another instance of Bob's dangerous overreliance on a group of terrorists over his personnel.

As a quick aside, most people do not understand that Bob was never intended to be the COB. In approximately December 2011, the actual COB of Benghazi was sent home early, and Bob was sent to fill the gap for about one month until the Near East Division at CIA Headquarters could choose a COB. Instead, Bob remained month after month, even as he had no experience leading in a warzone. Before coming to Benghazi, CIA Leadership already knew Bob could not perform as an effective leader in a warzone, but they put CIA employees at risk under him regardless. Bob only served in one warzone as a Case Officer, where he only did a partial tour as he had been embroiled in a scandal concerning the payoff of CIA assets after the death of an asset due to his poor tradecraft.

Now when it comes to these delays, the waits, the standdown, however one feels comfortable describing it, we need to make clear again: this was only due to our local CIA Leadership on the ground. The Director of the CIA, General Petraeus, and the COS in Tripoli were not even aware in real-time that we weren't permitted to go. In addition, the CIA would never need to wait for directives from the State Department or the Department of Defense to respond to a localized crisis.

About the Authors

Sarah Adams and Dave Benton worked together at the Central Intelligence Agency's Annex. This book is the sharing of their personal, cold-case investigation into the attacks that occurred between September 11th and 12th, 2012, in Benghazi, Libya.

SARAH ADAMS

Sarah was an Intelligence Analyst and Targeter with the Central Intelligence Agency (CIA). She also served as the Senior Advisor to the U.S. House of Representatives Select Committee on Benghazi. Currently, she leads Department of Defense (DoD) research and development efforts to discover, incubate, and deliver innovative data-driven, technology-enabled solutions to answer complex national security challenges.

DAVE BENTON

Dave has over 25 years of serving in specialized teams and leadership roles in the military, with law enforcement, in protective security, and was also part of the U.S. Government's Intelligence Community. He was a member of the Annex security team that responded to the September 11th, 2012, terrorist attacks on the U.S. Consulate in Benghazi. His story was told in the book "13 Hours: The Inside Account of What Really Happened in Benghazi," authored by Mitchell Zuckoff.

www.ingramcontent.com/pod-product-compliance
Lightning Source LLC
Chambersburg PA
CBHW030335010526
44119CB00028B/405/J